INFANT DEATH EXONERATIONS

THE FOLBIGG CASE

A PRACTICAL GUIDE FOR FAMILIES, ADVOCATES AND PROFESSIONALS

JOHN SYDNEY SMITH MD

First Edition

ISBN: 978-1-7644355-1-2

Cover design and layout by John Sydney Smith

Independently published through Amazon KDP

[www.johnsydneysmith.com

For permissions or more information, contact:

johnssmith@outlook.com.au

TABLE OF CONTENTS

PREFACE

This book investigates one of the most extraordinary criminal justice stories of modern times: the case of Kathleen Folbigg, whose wrongful conviction for the deaths of her four children became a global symbol of scientific, legal, and social fallibility.

Spanning decades and continents, the narrative draws together medical history, genetics, flawed forensic traditions, the evolution of innocence movements, and the lived trauma of wrongful incarceration. Through deeply researched case studies—anchored in the unfolding tragedy of the Folbigg family—this book explores how rare genetic variants, systemic bias, and faulty expert testimony converged to create and sustain a miscarriage of justice.

By situating Kathleen's ordeal within the broader context of child death investigations, gendered suspicion, and forensic reform, this work aims not only to chronicle one woman's journey from grief and vilification to exoneration but also to reveal how law and science must work together to prevent future injustices. Modern genetic discovery and global advocacy are now reshaping our understanding of both innocence and culpability.

It took two public inquiries before Kathleen was exonerated. The first—an adversarial proceeding—failed to understand or accept the extraordinary genetic abnormalities that underpinned the children's deaths, instead relying on outdated forensic assumptions and lay interpretations of her diaries. The second inquiry, by contrast, set a new international precedent: for the first time, a learned scientific body (the Australian Academy of Science) was empowered to recommend experts, guide the inquiry's scientific standards, and scrutinize how the justice system absorbs new knowledge. Breakthroughs in the genetics of cardiac arrhythmia and sudden death unfolded during the proceedings themselves, ultimately forcing a reappraisal of the evidence.

It is hoped that this melding of law and science will not remain exceptional but set a global standard—protecting families everywhere from the devastation of wrongful conviction and providing hope for those wrongfully incarcerated.

ACKNOWLEDGMENTS

In early 2022 I first became interested in sudden unexplained infant deaths, but particularly the plight of Kathleen Folbigg, who had been incarcerated since 2003 for the alleged suffocation of her four children. Her case had garnered a lot of media attention in 2019 when an inquiry rejected the evidence of a possible genetic cause for two of her children, instead relying on lay interpretations of her copious diaries to confirm her guilt.

On reading the transcripts of that inquiry and the documents relating to her 2003 trial, I concluded, as had others, that the verdicts were unsound. While the prosecution claimed at her trial that her firstborn, Caleb had only minor health problems prior to his death at the age of 19 days, there was strong evidence to the contrary. It was claimed that when the second born, Patrick, was first taken to hospital after losing consciousness he was suffering the effects of smothering, whereas there was ample evidence that he had just suffered an epileptic fit. The last born, Laura, was said to have been smothered, yet there was autopsy evidence of a significant inflammation of her heart and evidence that abnormalities in the recordings of her heart rate and breathing during sleep had been misinterpreted.

Concerned by this information I wrote to Isabel Reed, counsel for Kathleen, about Patrick and Laura and that started my journey. Isabel passed on my information to Rhanee Rego and Robert Cavanaugh who were then representing Kathleen. Graciously they accepted my input and put me in touch with David Wallace, who was the first to suggest to the team that genetic abnormalities might account for the children's deaths, A

and got Professor Carola Vinuesa involved in the sequencing of the genes of Kathleen and her four children. I am eternally grateful to Isabel, Rhanee, Robert, David and Carolla for involving me in the plight to free Kathleen, and especially thankful to Carola for tutoring me on genetics.

CHAPTER 1 – KATHLEEN'S STORY

"Although the world is full of suffering, it is also full of the overcoming of it."
– Helen Keller

Introduction

Kathleen Folbigg's life embodies tragedy, stigma, and, most recently, scientific vindication. Her case, now recognized globally as one of the greatest miscarriages of justice in Australian history, was shaped by personal trauma, the extraordinary loss of four children to unexplained deaths, and the limitations of forensic science at the time.[1] Understanding her childhood and family history is vital context for what followed: a multi-decade legal saga culminating in a 2023 pardon after two decades behind bars, spurred by new genetic evidence that finally explained the loss of her children.[2]

Taffy Britton

Kathleen's father, Thomas John Britton—known as Taffy—struggled with alcohol, rage, and relationships. Born in Wales, he served in the Royal Navy but jumped ship in Australia after WWII. He had relationships with local criminals and worked for infamous drug lord Robert Trimbole. In 1952, he was jailed for slitting the throat of his first wife and later, in 1969, received a life sentence for stabbing Kathleen's mother 24 times in a drunken, jealous rage.[3] After serving 14 years, he was deported.

Kathleen's Early Resilience

Kathleen Megan Britton was born on June 14, 1967, in Balmain, a suburb of Sydney, Australia. For extended periods she was left with her maternal aunt, Mrs. Platt, and her husband. On the day after her mother's murder, she was made a Ward of the State and placed in the care of the Platts, under the supervision of the Department of Community Services (DOCS). She was 18 months old.

Four months later the Platt's reported that Kathleen seemed to be intellectually dull as she had difficulty learning the basics of hygiene and manners, she was aggressive, and she engaged in sexualized behaviour. State doctors recorded that she may have been sexually abused by her father and her maternal grandmother was a "destructive influence."[4]

In June 1970, the Pratts declared that they could no longer care for Kathleen. She was moved to Bidura House, a government orphanage in Sydney. In July 1970, testing suggested that she had an IQ of 77, but she had not cooperated with the assessment. She was transferred to the Corelli Babies Home which mainly cared for disabled children.

Foster Care

When she was three years of age, Kathleen was placed in the foster care of Deidre and Neville Marlborough who lived Kotara, a suburb of Newcastle, New South Wales. Their 20- year-old son Russell, and their 17-year-old daughter Lea had left home, and they yearned to raise another child. As her foster father was preoccupied with his accounting business, the parenting was left to her foster mother. But she was emotionally aloof and instilled in Kathleeen the belief that feelings should not be displayed or discussed. She was also a strict disciplinarian and punished Kathleen for any perceived indiscretion by striking her with a belt or stick or slapping her around the head. Russell and Lea doted on Kathleen when they visited her, and they protected her when their mother was too demanding. Kathleen formed a close relationship with Lea.[4]

Kathleen settled into the local Kotara Primary School, although she was somewhat of a loner. In

Year 2, testing showed that she had an above average IQ of 110, and her teacher wrote that she was exhibiting leadership qualities. But in Year 5, a teacher described her as "inattentive, disruptive and defiant," and Kathleen and her foster parents attended a Community Health Centre for counselling.

In Year 6 Kathleen appeared before the Worimi Children's Court and was admonished after shoplifting some chocolates, but this was her only clash with the law. She also truanted. She repeated Year 6 so that she could improve her peer relationships. She then moved to the co-educational Kotara High School, successfully completed Year 11, and started Year 12, with average academic results.

In her early school years Kathleen had few friends and spent a lot of time alone in her room. From the age of 13 years, she was rebellious and argued more with her foster mother, especially over her freedom to mix socially. At the age of 14 she made enquiries about her origins through DOCS and first learnt that her father had murdered her mother. Her foster mother then used this as an emotional weapon, accusing Kathleen of sharing his temperament.

Kathleen was about 15 years of age when she suffered her first syncopal episode. This was witnessed by her friend and classmate Billy-Jo Ann

Buckley. They were taking part in a school swimming carnival when it occurred. Billy-Jo later wrote:

> *Kathy and I started off the race together, but I got ahead of her in the pool. I swam the length of the pool, which I recall was 25 metres. The race was only one lap.*
>
> *When I completed the lap and got out of the pool. I saw people were with Kathy on the left-hand side of the pool. Kathy was lying down and struggling to get up. I was not sure what was going on but knew they were attending to her.*
>
> *I did not speak to Kathy at this time. I was with our friends Tracy Chapman and Megan Donegan. I spoke to them about what had happened to Kathy, and they said she had a panic attack and had to be retrieved from the bottom of the pool.*

Aged 15 Kathleen started a relationship with a fellow student, much to her foster mother's chagrin. She established friendships through his peer group, and she still has contact with some. At the age of 16 she was described as a sensible, well-adjusted young woman who enjoyed life. She adopted her foster parents' surname. She found a casual job in a service station.

Kathleen's Independence

When Katheen was 17-and-a-half her foster mother again tried to slap her across the face, but Kathleen revolted. She blocked the blow, packed her bags and left home to live with Billy-Jo and her mother. She discontinued her Year 12 studies and found a job working as a waitress in an Indian restaurant.

Shortly after her move she had her second observed syncopal episode. Billy-Jo wrote that Kathleen went to the home bathroom:

> *When Kathy came out of the bathroom she said words to the effect "I do not feel well" and I noticed that her face was pale. As soon as she said that she collapsed and fell onto the floor in the hallway outside of the bathroom. She was unconscious for approximately 30 seconds. When Kathy regained consciousness, she was disorientated and said she did not feel well. I was not sure what to do so I called my local GP … The GP told me to bring Kathy down to his surgery as soon as I could.*[5]

Kathleen enjoyed dancing and it was on the dance floor, when she was 18, that she met . He was six years older. She was impressed by his charm and outgoing nature and enjoyed his compliments on her good looks. He seemed stable, as he had a good job in a clerical role with BHP Steel. His mother died during childbirth when he was 15, but he was close to his four brothers and three sisters. Shortly after they met, they set up home in a rented flat in Georgetown in Newcastle. When she was 19, they became engaged and purchased a home in the Newcastle suburb of Mayfield.

Summary

Kathleen Folbigg's background is marked by instability, trauma, resilience, and brief glimpses of normalcy and support. Her formative years—from bereavement and abuse to academic highs, family turbulence, and young adulthood—shaped the woman later thrown into the crucible of public suspicion and scientific ambiguity. Understanding the context of her early life is critical to appreciating the immense personal challenges she faced amid, and after, the deaths of her four children—a subject examined in the next chapter.

CHAPTER 2 – MOTHERHOOD

"Unable are the loved to die. For love is immortality."

– Emily Dickinson

Introduction

This chapter examines the devastating course of Kathleen Folbigg's life as a young mother: the birth and sudden deaths of four children under suspiciously similar circumstances, the unraveling of her marriage under extraordinary pressure and grief, and the role her private diaries played in turning grief into suspicion.[1] Centering on lived experience, marital tension, and evolving investigation, this account foreshadows the scientific controversy and eventual exoneration that would follow.[2]

Four Tragic Deaths

On September 5, 1987, Kathleen married Craig Folbigg. Despite his initial support, Kathleen's foster parents withheld warmth and approval, her foster father refusing to walk her down the aisle and her mother sitting at the back of the church. Another episode in a lifetime of rejections, Kathleen learned early to hide pain behind self-control.[3]

Some seven months later, Kathleen became pregnant. Determined to provide their child with a healthy environment, Craig promised not to smoke at home or around future children.

Caleb Folbigg

Kathleen's pregnancy with Caleb was mostly uneventful but marked by two episodes where she lost consciousness; medical assessment blamed these on syncope rather than epilepsy or underlying cardiac disease. At 38 weeks, a further episode led to brief hospitalisation. Caleb was born on February 1, 1989. Labor was prolonged and complicated—he was delivered using Kielland's forceps, necessitated by abnormal fetal position and distress. Research papers scanning the 1979 to 1999 period indicated that when these had to be used, the infant had a 3% chance of dying, a 10-15% risk of suffering neurological problems, a 23% risk of cerebral irritation, and a 17% chance of having breathing difficulties.[4]

Caleb showed classic signs of Respiratory Distress Syndrome and mild pneumomediastinum—a leak of air into the space around the heart. For the first days, he required supplemental oxygen and careful monitoring. He also exhibited abnormal feeding, needing to pause between sucks to catch his breath, which a pediatrician attributed to laryngomalacia—a "floppy" larynx that can cause airway instability, feeding problems, and, in up to a quarter of cases, be life-threatening.[5]

Kathleen, diligent and anxious, meticulously tracked every feed, symptom, and routine through her diaries—private reflections that would later be misread as sinister confessions. Caleb's bassinet was in a sunroom beside their bedroom; Craig, a heavy sleeper with undiagnosed sleep apnea, offered little assistance during night feeds. Kathleen's efforts were solitary and exhaustive.

On the night of February 20, after fussiness and "restlessness" with feeding, Caleb was put to sleep in his basinet. At 2:50 am, Kathleen found him unresponsive, blue-lipped, and cold. Despite frantic CPR by Craig and immediate ambulance response, resuscitation failed. Postmortem examination found no external trauma and concluded "cot death" (SIDS) as cause, noting that the possibility of laryngomalacia could not be confirmed.[6]

Grief fractured the couple's fragile bond. Craig retreated into stoicism, Kathleen into distraction and isolation. They coped separately with numbness and guilt, attending grief counseling and genetic risk assessment. Kathleen's diary and Craig's occasional written letters reflect their struggle to find meaning in loss.

Patrick Folbigg

By September 1990, Kathleen was again pregnant. Patrick was born full term on June 3, 1990, after another complicated labor involving cord and airway issues. Within weeks, the family became familiar visitors at hospital outpatient clinics for sleep studies and developmental checks; Patrick's early health issues included transient torticollis and feeding difficulties, but nothing immediately alarming until the sleep crisis of October 18.[7]

On that morning, Kathleen found Patrick unresponsive, blue, and gasping. Paramedics found him "hard to wake," pale, listless. At the hospital he recovered rapidly after oxygen, and it was recorded that he was "awake playing and responding to parents appropriately." Initial EEG and brain ultrasound tests were normal, but the medical staff questioned whether he was recovering from a seizure

. This episode was later classified as an Acute Life-Threatening Event (ALTE).

Within 40 hours, however, Patrick suffered his first witnessed major epileptic seizure—a three-minute convulsion followed by partial seizures and a rapid decline. Over the next months, he was diagnosed with epileptic encephalopathy with progressive loss of vision, likely due to brain damage from repeated fits.

Across multiple hospitalizations, with signs that raised the possibility of a genetic or metabolic disorder, Kathleen remained an attentive and devoted carer, facilitating therapy and tracking every aspect of his decline. On February 13, 1991, she found Patrick unresponsive. Resuscitation failed, and Dr. Wilkinson certified death due to asphyxia with underlying severe neurological disorder. Forensic investigation found no trauma; cause remained "natural but unknown."[8]

Devastated and guilt-ridden, Kathleen blamed herself, questioning her ability as a mother and suspecting she had passed on a genetic defect. Her grief—deepened by marital neglect and a sense of isolation—manifested in depression, overeating, and diary confessions of despair.

The first seeds of suspicion and resentment grew after Kathleen discovered Craig embracing her best friend. The couple sold their house and moved to Maitland, but cracks in the marriage deepened amid mutual blame and emotional withdrawal.

In late 1991, they decided to have another baby. To be in better health for the pregnancy Kathleen went on a diet and exercise regime.

In February 1992, Kathleen and Craig were told by Dr. Alison Colley, geneticist, that there was possibly a one-in-four chance that another child could die from some yet unknown genetic disorder, but it was possible that this only affected the male offspring.[9]

Sarah Folbigg

Sarah was born on October 14, 1992. Sarah was healthy, attended playgroups, and was monitored at home with an apnea alarm mat. After normal pediatric reviews and a clear sleep study conducted at Kathleen's request, the monitor was eventually discontinued.

In August 1993, after days of croupy symptoms, Sarah died suddenly in her sleep even as her mother checked on her repeatedly. Both parents attempted CPR prior to ambulance arrival; an autopsy found a highly unusual obstruction by the uvula and widespread inflammation of the upper airway, establishing SIDS as the official cause but leaving "natural causes cannot be ruled out" in medical notes.

The third unexplained loss devastated the couple. Mutual support was fleeting. Craig sought counseling; Kathleen retreated into guilt, depression, and obsessive self-blame reflected in the diaries, which she increasingly used as a sole outlet for distress.

Marital Breakdown

Their double grief drove deeper wedges. Craig's work kept him away, and strains multiplied with financial worry and the echo of blame. Kathleen, newly aware of Craig's infidelities (discovered as she contracted a venereal infection), recorded her anger, humiliation, and loss of self-esteem in her diary. At times separated, often sleeping apart, the marriage was held together mainly by inertia and societal expectation.[10]

In late 1996, to salvage the relationship, they decided to try for another child. Kathleen's next pregnancy was uneventful, and she took pains to improve her health for Laura's birth.

Laura Folbigg

Laura was born healthy in August 1997, but soon sleep studies revealed moderately severe "central apnea"—breathing pauses due to brain signaling problems. Home monitoring showed multiple periods of abnormal heart rate and episodes of apnea, but these key abnormalities were missed by technicians and never reviewed by doctors.

Kathleen remained frustrated by Craig's lack of support.[11] Medical specialists later testified that abnormal home monitor readings (with heart rates as low as 24 bpm) should have prompted further genetic and cardiac investigation. In one frightening incident, Laura stopped breathing while being babysat but revived spontaneously. Laura's medical history, sleep studies, and recurring viral illnesses formed a backdrop to parental anxiety and relentless vigilance.

On March 1, 1999, after several days of illness, Kathleen found Laura unresponsive in bed around noon. Despite CPR and rapid ambulance response, Laura could not be revived. Autopsy listed inflammatory myocarditis (heart muscle inflammation), but at trial, forensic pathologists contested if this was sufficient to explain her death—witnesses at the 2022-23 review would confirm this myocarditis was indeed "lethal grade."

Diaries and the Realities of Motherhood

Kathleen's diaries were extensive, spanning from the early 1990s through the years of her family trauma. Approximately one-third of her entries discussed her marital problems. For example, in mid-1996 she wrote about a genital infection, adding: "It is supposedly sexually transmitted, by Craig to me." In September 1996, she described Craig as controlling, stating: "Can't stand it when Craig talks down to me like I'm some sort of inferior being." Later, she confided:

> *"I think I wish to party because I'll be pregnant soon and know once that happens, he won't let me out of the house again. Even consider not telling him straight away?"*

By October 1996, her frustration deepened:

> *"Craig's being a childish pratt again. Sometimes I wonder why I'm bothering. The thought of returning on the pill and forgetting about having a baby with him is attractive. His*

childish behaviour about me going out is really dragging me down."

There were also several entries that the prosecution claimed were admissions of guilt in the deaths of her children:

- On the day Caleb died, she wrote at 1:00 am that he was "a little restless" and at 2:00 am, "finally asleep."
- Of Sarah, she reflected: "I knew I was short tempered and cruel sometimes to her and she left. With a bit of help."
- When pregnant with Laura, she agonized over her past struggles and wrote:

  "Another year gone & what a year to come. I have a baby on the way … This time I am going to call for help this time & not attempt to do everything myself any more – I know that that was the main reason for all my stress before & stress made me do terrible things..."

- In an entry discussing Laura, Kathleen wrote: "I was depressed, angry and upset after I lost it with Laura."
- Of Caleb and Patrick, she confided: "I often regret Caleb and Patrick, only because your life changes so much and, maybe, I'm not a person who likes change."

While these lines were portrayed in court as confessions or signs of dangerous intent, a bevy of later experts—psychologists, psychiatrists, and grief specialists—explained that mothers suffering overwhelming stress, depression, and trauma after losing a child often describe themselves as failures and may reinterpret ordinary frustration or negative emotion as "terrible." None of these entries, they maintained, could reasonably be called admissions of murder; instead, they revealed guilt, anxiety, and self-blame common to parents coping with devastating loss.

Turning Grief into Guilt

After Laura's death, Kathleen's relationship with Craig fully disintegrated. Fights were punctuated by periods of estrangement and attempted reconciliation. Kathleen returned to work, struggled with depression, and wrote increasingly desperate diary entries—acknowledging her guilt, sadness, and a sense of abandonment from friends and family.

When Craig found her diary after their final separation, he delivered it to police, marking the start of her prosecution. The diary—full of ambiguous self-blame, sadness, and confusion—was characterized by the prosecution as a series of confessions, cementing the case against her despite expert psychiatric testimony that the entries reflected trauma and maternal grief, not criminal intent.

A New Relationship

In August 2000, Kathleen commenced a relationship with Tony Lambkin. Lamkin realized that something was troubling Kathleen as sometimes she cried all night. A few months into the relationship Kathleen suddenly announced, "Tony there's something hanging over my head … I need to talk to you about it … I've been accused of killing my four children." She related the history and explained that Craig had turned against her once their separation became permanent. Lambkin believed in her innocence and the relationship blossomed. He observed her happy interaction with children and reported, "It melted my heart. It also confirmed that she was the woman whom I one day wanted to be the mother of my children."

Summary

Four medically complex and ultimately unexplained child deaths shattered Kathleen Folbigg's life, unraveling her marriage and leaving her as both victim and, in the eyes of the legal system, the principal suspect. Compounded by marital strife, ambiguous diary entries, and the prosecution's interpretation of inner grief as evidence of guilt, the Folbigg case stands as a pivotal lesson in the interplay of tragedy, suspicion, and misunderstanding. The chapters ahead will explore in depth the clinical science of sudden child deaths, the anatomy and genetics central to these cases—and the broader medical, legal, and gendered context that led to Australia's most infamous wrongful conviction.

CHAPTER 3 – CHILDHOOD DEATHS IN HISTORY

"Let us reach out to the children. Let us do whatever we can to support their fight to rise above their pain and suffering."
— Nelson Mandela

Introduction

Childhood deaths were tragically common throughout human history, shaping societies and family structures until major improvements in public health emerged in the twentieth century. This chapter examines historical childhood mortality, medical misbeliefs—and the infamous Doudet affair—as pivotal events in the emergence of child protection laws and understanding sudden unexplained deaths in infants and children.

Childhood Mortality

Archaeological and historical research indicates that for over 12,000 years, nearly one-quarter of all children died before reaching their first birthday, and half before adulthood.[1] Malnutrition led to diseases such as rickets, while overcrowding, poor sanitation, and contaminated water made gastrointestinal infections and tuberculosis common.[2] Plagues intermittently devastated communities, and infanticide was condoned in some societies until it was outlawed in Europe and the Near East during the first millennium.[3]

Sudden Unexplained Deaths

Throughout history, infants and children sometimes died suddenly and inexplicably, often during sleep. The first recorded account appeared in the diary of Massachusetts judge Samuel Sewall:

> *February 13, 1686. Mr. Eyre's little son dyed, well to bed: dyed by them [the parents] in the Bed. It seems there is no Symptom of Over-laying.[4]*

On March 9, 1863, a more detailed description emerged, hinting at genetic predisposition, when Dr. David Cheever presented to the Boston Society for Medical Improvement:

> *An infant, 10 weeks old, apparently in perfect health, suddenly died while sleeping... An infant*

cousin of the child had died in a precisely similar manner.[5]

In the 19th century, sudden infant deaths were commonly blamed on disease of the thymus gland. The thymus, part of the lymphatic system, plays a key role in immune defense.[6] In 1830, Dr. J.H. Kopp suggested that an enlarged thymus caused suffocation ("thymic asthma"), but no obstruction could be shown.[7] In 1889, Professor A. Paltauf proposed that pathology affecting the entire lymphatic system rendered infants prone to collapse—a condition called "status thymo-lymphaticus."[8] These hypotheses merged, and clinicians attempted to treat this supposed syndrome by surgically removing or irradiating the gland, leading to disastrous outcomes.[9] Surgical removal carried mortality rates up to 30%.[10] Irradiation with x-rays increased thyroid cancer risk a hundredfold.[11]

By 1931, British investigators concluded that there was no such condition as status lymphaticus.[12] Tragically, these misguided interventions were based on flawed normal standards and poor imaging methods, and autopsies mostly on severely malnourished children from poor families whose thymus was atrophied led to erroneous conclusions.[13]

Child Abuse

In 19th century British middle-class households, the father was the breadwinner and decision-maker; the mother managed the home. Children were considered parental property and lacked government protection. Sexual matters were taboo, opposed by religious leaders, and masturbation was seen as a grave moral evil causing blindness, insanity, or premature death. Treatments included restrictive garments, surgical procedures, and severe discipline for transgressions.

Child discipline was harsh and often corporal, rarely recognized as cruelty as it was regarded as a parental prerogative—until the "Doudet affair" changed perceptions.[14]

The Doudet Affair

Flore Marguerite Celestine Doudet (b. 1817, Rouen) was educated in Paris before serving as governess to royalty and nobility. In March 1852, she became governess to the five children of widowed Dr. James Loftus Marsden—Lucy (13), Emily (12), Mary Ann (11), James (9), and Rosa (8).[15] Marsden practiced homeopathy and hydrotherapy, imposed strict discipline and a protestant ethic, and treated any sexual misdemeanor—such as masturbation—with severe punishment.

Soon after hiring Doudet, Rosa was accused of theft and punished by being locked in a room for three days. Later, Doudet alleged that all the Marsden daughters engaged in immoral behavior, leading to more punishments and the employment of physical restraints (hands and feet tied at night). In June 1852, Doudet took the girls to her Parisian apartment, where the abuse escalated: starvation, confinement, beatings, hair-pulling, and writing forced confessions.

Concerns arose among neighbors and even a consulting homeopath, but explanations for the girls' emaciation were attributed to "uncontrolled onanism" (masturbation). On May 24, 1853, Doudet struck Mary Ann, resulting in paralysis and subsequent death two months later. Marsden moved his surviving daughters to their aunt Fanny, and Lucy soon died, her health never recovering from the ordeal.

Marsden reluctantly pressed charges against Doudet. The trial for manslaughter commenced on February 21, 1855. Servants testified that there was no evidence of immorality, but the defense portrayed the girls as manipulative and Marsden as violent. The jury split evenly; Doudet was acquitted of manslaughter but later convicted of cruelty and imprisoned for two years, later increased to five on appeal, though she was pardoned by Napoleon III in 1858.

The affair's broad publicity forced recognition that cruelty could occur in seemingly respectable households and inspired reforms in child protection.

Reform and Recognition of Child Cruelty

Auguste Ambroise Tardieu led the autopsy on Mary Ann Marsden and was a pioneering forensic scientist and reformer. He documented child abuse and unsafe child labor, most notably in his 1860 book on cruelty to children ("Etude medico-legale sur les sevices et mauvais traitements exerces sur des enfants").[16] Tardieu's work described the battered child syndrome, exposed infanticide and non-accidental trauma, and emphasized that parents and guardians could be perpetrators—a radical notion for his era.

Tardieu also hypothesized that spots on the thymus signified smothering, but this was later disproved.[17] The belief that overlaying (parents accidentally smothering infants while sleeping) was a cause of sudden deaths was prevalent, but separating infants from parents did not reduce mortality rates.[18]

Mary Ellen Wilson and Early Child Protection

Physical abuse of children in 19th-century America was widespread until the highly publicized case of Mary Ellen Wilson. Orphaned and fostered out, she was starved, beaten, and confined for years. Social worker Etta Wheeler and activist Henry Bergh (founder of ASPCA) intervened when authorities would not, leading to the first successful prosecution for child abuse in the US in 1874.[19]

Mary Ellen's case pioneered organized child protection: Bergh founded the New York Society for the Prevention of Cruelty to Children, and new laws were enacted.[20] Mary Ellen overcame her traumatic childhood, married, raised her own children, and lived to age 92, her story symbolizing the beginning of child welfare movements.[21]

Summary

For millennia, childhood deaths were the norm, rooted in poverty, disease, and misunderstanding. Only through shocking cases like the Doudet affair and the advocacy of reformers such as Tardieu and Bergh did modern societies recognize child abuse and advance public health and protection laws. This chapter reveals the interplay of social attitudes, flawed medical theories, and gradual reform that shaped contemporary understandings of childhood mortality and mistreatment.

CHAPTER 4 – SUDDEN UNEXPLAINED DEATHS

"There is no place for dogma in science. The scientist is free, and must be free to ask any question, to doubt any assertion, to seek for any evidence, to correct any errors."
— J. Robert Oppenheimer

Introduction

This chapter traces the history of unexplained childhood deaths, focusing on sudden infant death syndrome (SIDS) and its shifting scientific, social, and legal interpretation throughout the twentieth century. It highlights the evolving research, debates, and tragic consequences for families, especially as faulty scientific evidence at times led to miscarriages of justice.

Unexplained Sudden Deaths

In the early twentieth century, global child mortality rates began to decline substantially, halving by 1950 due to improved sanitation, nutrition, vaccination, and antibiotics.[1] However, unexplained sudden deaths in infants—what would become known as SIDS—remained a persistent and distressing phenomenon. Researchers were perplexed by its ongoing prevalence even as infectious causes receded.

Initially, theories about SIDS clustered around sleep position. Pathologist Harold Abramson reported in 1944 that most unexplained infant deaths in New York City occurred while infants slept prone (on their stomachs); he suggested developmental immaturity prevented babies from moving away from suffocating bedding and recommended discontinuing the prone position.[2] Edward Woolley, a pediatrician in the UK, shortly after argued that bedding rarely caused dangerous oxygen drops and instead pointed toward infections or choking, yet this debate highlighted ongoing uncertainty.[3] By the 1950s, authoritative medical texts gave inconsistent advice; the influential Mitchell-Nelson Textbook of Pediatrics offered little clarity, and Dr. Benjamin Spock shifted recommendations in different editions.[4]

Interest in understanding these deaths slowly increased due to parental advocacy. When the Roe family lost their son Mark to unexplained infant death, the Mark Addison Roe Foundation began funding research and support in the 1960s. Government agencies and international conferences gradually recognized the need for systematic research, eventually leading to formal recognition and definition of "sudden infant death syndrome" in 1969.[5]

The experience of Judith Choate, whose grief was compounded by suspicion and insensitive questioning from police and medical officials, exposed the social costs of unexplained deaths and galvanized families to advocate for research and sensitive bereavement support. Organizations such as the National SIDS Foundation in the U.S. and its international counterparts were formed in the following decades.[6]

Prone Sleeping a Factor

By the 1970s, studies increasingly found strong statistical links between prone sleeping and SIDS.[7] Yet, despite growing evidence and repeated calls for change, actual sleeping position practices remained slow to evolve—in part due to physician uncertainty, lagging public health messaging, and entrenched advice. It is estimated that tens of thousands of preventable deaths occurred in the ensuing decades.

A powerful 1985 study out of Tasmania—and a prospective survey involving over 11,000 infants by Terry Dwyer and the Menzies Institute—corroborated that prone sleeping (especially when combined with other risk factors like soft bedding, room heating, or swaddling) was a major cause of SIDS deaths.[8] Public health campaigns such as Australia's "Reducing the Risks of Cot Death" (1991), the UK's "Back to Sleep" (1991), and the U.S. "Back to Sleep" campaign (1994) quickly resulted in dramatic reductions in SIDS mortality. International meta-analyses still confirm that the risk of SIDS is around 3–7 times higher for infants sleeping prone compared to supine.[9] [10] [11]

A leading theory, the Triple Risk Hypothesis, posited that SIDS occurs where (1) an infant is intrinsically vulnerable, (2) during a critical developmental period, and (3) when faced with an external stressor such as infection or overheating.[12]

The Apnoea Theory

For years, it was also widely believed that episodes of apnea (cessation of breathing) caused SIDS, a claim advanced by Alfred Steinschneider's influential but later discredited 1972 study.[13] Critics warned of uninvestigated maternal roles and flawed methodology. Prosecutors and researchers later revealed that Steinschneider's work was at best erroneous and at worst fraudulent, with the Hoyt family case unmasking murders that had been misattributed to SIDS.[14]

The broader question of intentional suffocation or other forms of child abuse arose in the 1990s. David Southall's covert video surveillance work uncovered dozens of cases where caregivers were seen abusing or smothering infants.[15]

Around the same time Sir Roy Meadow introduced the controversial concept of "Munchausen Syndrome by Proxy," describing parents who deliberately harm their children to gain attention or sympathy. He dismissed the possibility that some cases of SIDS could have a genetic basis and formulated "Meadow's Law," which asserts: "one sudden infant death is a tragedy, two is suspicious, and three is murder until proven otherwise." Meadow calculated that as the risk of SIDS occurring in a family was 1/1,000, the probability of two natural infant deaths in one family was 1/1,000 by 1/1,000 and so one in a million, based on the assumption that such deaths are independent events and genetics play no role.[16] His statistical reasoning and methodology were later discredited, leading to several high-profile exonerations, including those of Angela Cannings and Donna Anthony.[17]

Genetic research revealed that familial recurrence of SIDS—especially among twins or in certain populations—was far higher than previously recognized, supporting a genetic hypothesis for some cases.[18] Landmark epidemiological studies demonstrated that subsequent SIDS deaths in a family were far from a one-in-a-million event; rather, the risk was three-to-six-fold increased compared to the general population.[19] As shown in Appendix II, this was reflected in 18 studies published between 1971 and 2000 that reported on 254 families in which recurrence occurred, affecting two infants in 231famuilies, three in 16 families, four in 4 families, five in one family and six in another family. Legal and scientific consensus gradually shifted: recurrence alone was deemed insufficient evidence of murder in the absence of additional compelling proof.[20]

Summary

Sudden unexplained child deaths remain a complex interaction of biological, environmental, and social factors. Changing scientific consensus, advocacy by bereaved families, and evolving public health campaigns have dramatically reduced SIDS incidence. However, flawed science and "laws" based on debunked statistics irreparably harmed families. Ongoing research continues to unravel both genetic susceptibilities and societal factors, and the legacy of legal and medical errors has shaped how cases are investigated, prosecuted, and understood today.

CHAPTER 5 – THE SHAKEN BABY SYNDROME

"The greatest obstacle to progress is not ignorance, but the illusion of knowledge."
— Daniel J. Boorstin

Introduction

Shaken Baby Syndrome (SBS) emerged in the late twentieth century as a medical diagnosis and forensic doctrine used to explain and prosecute cases of sudden, unexplained infant brain injury and death. This chapter examines the origins of the SBS hypothesis, its propagation through medical, legal, and public arenas, critical scientific reevaluation, and the resulting controversies, wrongful convictions, and reforms.

The Origins and Early Development of SBS

John Caffey, an American paediatric radiologist, identified a correlation between subdural hematomas (SDH) and long bone fractures in infants as early as 1946. He could not explain the association and noted in some cases the infants were unwanted.[1] Charles Henry Kempe, in 1962, introduced the concept of "battered child syndrome," arguing that serious injuries in children often originated from parental or guardian abuse.[2]

In 1971, British neurosurgeon Arthur Norman Guthkelch proposed that SDH in infants could be caused by whiplash injuries sustained from violent shaking.[3] In support, he referenced a 1968 experimental study by Ayub Ommaya, who used rhesus monkeys to show that whiplash injury could cause subdural hemorrhage and brain damage, but only when the forces involved were extreme and typically produced neck fractures.[4] Ommaya warned Guthkelch directly that the forces generated by shaking an infant were insufficient to cause SDH without concomitant traumatic injury; however, this caution was not emphasized in Guthkelch's publication, inadvertently paving the way for Caffey's popularization of the SBS hypothesis. Caffey, in 1972 and 1974, hypothesized that shaking could produce the constellation of SDH and skeletal fractures despite conceding the evidence was incomplete and circumstantial.[5][6]

The term "Shaken Baby Syndrome" was coined in 1984 by Ludwig and Warman, who defined the triad of diagnostic features: encephalopathy (brain dysfunction), subdural hemorrhage, and retinal hemorrhages—claiming the presence of these triad features was diagnostic of SBS.[7]

Proponents then argued that the triad resulted only from shaking, the effects are immediate, and the last person with the child is responsible. This *res ipsa loquitur* doctrine made SBS a near-immutable diagnosis of abuse or murder, facilitating convictions often in the absence of motive or prior abuse.

Rise and Challenge of SBS Diagnoses

SBS became a staple in paediatric, forensic, and legal spaces. Hospitals, legal systems, and public health authorities were trained to recognize and act on the triad, with punitive measures. Doctors who questioned or defended accused parents risked professional censure. Meanwhile, infants who died suddenly and would have previously been labelled as SIDS were increasingly investigated for SBS.

Peer-reviewed scientific work began to undermine the claims made for SBS. In 1987, Anne-Christine Duhaime and colleagues demonstrated in biomechanical studies that impacts, not shaking, generated forces capable of causing SDH and brain injury. Vigorous shaking produced forces far lower than a short fall.[8]

The Legal and Scientific Backlash

Concerns mounted that criminal trials for alleged SBS were predisposed toward conviction. Judge Abner Mikva warned in 1990 that child abuse trials gravitated toward guilty verdicts regardless of evidence.[9] Advocacy materials, such as the UK's "Handle with Care" leaflet in 1995, continued to promote shaking as a proven cause, despite mounting contrary scientific evidence.[10]

John Plunkett, a Minnesota forensic pathologist, began challenging SBS claims after reviewing fatal pediatric head injuries from short distance falls, concluding that the SBS triad could result from non-abusive mechanisms.[11] Geddes et al. published a series of influential papers (Geddes I, II, and III) in the early 2000s showing that the triad often reflected hypoxic insults, brainstem or cranial nerve injury, and dural bleeding unrelated to traumatic tearing.[12] Subsequent legal cases silenced, attacked, or prosecuted scientists who questioned SBS, including Geddes, Squire, and Plunkett.[13]

Professional Attacks and Persistence of SBS Doctrine

In the US, the term SBS was gradually replaced with "abusive head trauma" (AHT) by the American Academy of Pediatrics in 2009, broadening its scope but retaining the symptom triad as diagnostic criteria.[14] However, peer-reviewed research, conferences, and legal advocacy continued to expose alternative explanations for the triad: accidental trauma, congenital malformations, metabolic disorders, hypoxia, genetic conditions, and other non-abusive causes.[15]

Despite mounting evidence, legal systems remained resistant. High-profile exonerations, such as Jennifer Del Prete, Teresa Engberg, Brandy Briggs, Donna Conley, Letha Hockersmith, and Warren Hales in the US and appeals in the UK, exposed wrongful convictions based on SBS doctrine and highlighted flawed gatekeeping on scientific testimony.[16] Books and open letters called for courts to exercise caution when scientific consensus was lacking or expert disagreement was prominent.[17]

A timeline table of events in the SBS saga is presented in Appendix I.

The Bottom Line

To date, there has been no documented case in the medical literature of a baby developing the classical "shaken baby syndrome" triad (encephalopathy, subdural haemorrhage, and retinal haemorrhages) as a result of witnessed shaking alone. By contrast, multiple video-documented short falls have resulted in this triad.[18] [19] [20]

Summary

Shaken Baby Syndrome, once a dominant narrative in child abuse diagnostics, became a flashpoint for scientific, legal, and ethical controversy. Early theories, shaped by circumstantial evidence and clinical observation, provoked broad reforms in child protection but were later undermined by scientific research into traumatic brain injuries and hypoxic mechanisms. Despite years of advocacy and conviction, growing awareness of natural, accidental, and alternative causes led to critical reevaluation and the overturning of convictions. The history of SBS exemplifies the dangers of dogmatic forensic science, demonstrating the need for evidence-based medicine and rigorous legal standards before branding parents and caregivers as murderers.

HAPTER 6 – THE SCIENCE

"The child is not a miniature adult, and his diseases are not miniature adult diseases."
— Sir James Spence, British pediatrician

Introduction

By the late twentieth century, researchers undertook rigorous scrutiny of the scientific claims underpinning sudden unexplained deaths in infants and children. Statistical, anatomical, and genetic investigations have revealed a complex interplay of biology, environment, and inherited predispositions, challenging legacy explanations in apnoea, recurrence, and Shaken Baby Syndrome (SBS).

Infant Anatomy and Physiology

Infants possess unique anatomical and physiological characteristics that increase their vulnerability to fatal outcomes.[1] Their upper airway is crowded—the tongue and epiglottis occupy proportionally greater space—so obstruction readily occurs. Small, flexible tracheal rings, underdeveloped intercostal muscles, and a relatively crowded abdomen further restrict effective respiration. Infants compensate with rapid breathing rates: up to 60 breaths per minute in the neonatal period, slowing with age.

Chest wall compliance and immature musculature limit airflow, and a smaller bronchial angle allows easier aspiration of fluids. Dehydration (due to a high surface-area-to-volume ratio and underdeveloped fluid reserves) and hypothermia are common risks, compounded by immature thermoregulation. Similarly, limited hepatic and muscular glycogen stores—despite elevated energy needs—predispose infants to hypoglycemia.[2]

Neurological immaturity renders infants susceptible to seizures and periodic breathing (>75% of healthy neonates show intermittent breathing pauses). Birth itself greatly affects cerebral blood flow: nearly half of all full-term infants show subdural bleeds following delivery, especially with forceps or vacuum assistance. Most resolve spontaneously, however, subsequent physical stress such as a head injury, or hypoxia can trigger catastrophic events.[3]

Retinal hemorrhages also occur frequently in neonates (up to 50% with instrumental delivery), undermining SBS claims that these findings are uniquely diagnostic for inflicted injury.[4]

Childhood Epilepsy and Sudden Death

Epilepsy is the most common neurological disorder in childhood, affecting 0.5–1% of children and causing a minority of SIDS and unexplained deaths.[5] Research has shown that an unexplained ALTE (Acute Life-Threatening Event) can sometimes be the first sign of epilepsy, with incidence ranging from 4–13% of those later diagnosed with epilepsy, depending on cohort.[6] Videotaped deaths and registry studies have recorded epilepsy as a prelude to sudden death in children over one year of age, often accompanied by pathology in the hippocampal structures within the temporal lobes similar to SIDS cases.

Genetics and Disease

Mendel's discovery of dominant and recessive inheritance laid the foundation for understanding familial disorders.[7] Investigators later traced Huntington's disease and cystic fibrosis to single-gene disorders—dominant and recessive, respectively.[8] Human DNA, housed in 23 chromosome pairs, codes for 20,000–25,000 genes; molecular techniques now permit rapid, cheap sequencing of entire genomes.[9]

Essential amino acids (nine, sourced from diet) and non-essential amino acids (eleven, synthesized in the body) form the basis of protein structure[3]. Proper folding of the proteins, mediated by chaperone proteins, is vital for function. Modern sequencing (from

Sanger's original viral genome sequencing to nanopore-based bedside diagnostics) has transformed medicine, enabling genetic analysis of critically ill newborns in hours.[10]

Genetic databases (ClinVar, gnomAD) and standardized variant classification systems (ACMG) now underpin genetic diagnoses, categorizing them as pathogenic, likely pathogenic, benign, and VUS (variants of uncertain significance) by functional, population, and familial evidence.[11] Genes subject to "constraint"—where loss-of-function variants are incompatible with life—play a central role in sudden death syndromes.

A genetic variant is a change or difference in the DNA sequence of a gene compared to the typical (or "reference") version. These variants are often named using a letter-and-number code—for example, N54I means that at position 54 of a specific protein, the usual amino acid (asparagine, "N") is replaced by another (isoleucine, "I"). Such changes can be harmless, or they can alter how the gene or protein works—sometimes leading to disease or affecting risk for certain medical conditions.

CALM Genes and Sudden Cardiac Death

The CALM genes (CALM1, CALM2, CALM3) encode calmodulin, a key calcium-binding protein within the cells of the body. CALM variants are among the most constrained in nature, with only a fraction of possible mutations found in populations. These genes regulate cardiac ion channels and have been implicated in catastrophic arrhythmias in infants and young children.[12]

The QT interval measured on an electrocardiogram (ECG) is the time taken for the ventricle to contract to pump oxygenated blood out to the body and then fully relax before the next heartbeat. This interval can be abnormally long (termed the long QT syndrome -LQTS) or abnormally short (SQTS). Mutations can arise de novo (occurring in an egg or sperm cell of one of the parents, or directly in the fertilized egg) or be inherited, sometimes producing classic cardiac syndromes—including LQTS, SQTS and Catecholaminergic Polymorphic Ventricular Tachycardia (CPVT)—or more subtle arrhythmias and syncope.[13] Large international registries have shown that CALM variants present early (median onset 1.5–6 years), often manifest as sudden cardiac events, and may result in death or require device implantation by childhood.[14]

Recent studies identified the N54I mutation in CALM1 (and other CALM gene variants) as causative for severe, sometimes lethal arrhythmia in Swedish, Japanese, and global cohorts.[15] Expressivity is variable: some carriers remain asymptomatic into adulthood, while others succumb in infancy or childhood. Modern therapies include beta-blockers and implanted cardioverter-defibrillators, though gene editing promises future intervention.[16]

Inherited Heart Disease: Electrical Dysfunction and Arrhythmia

The heart's electrical system comprises sodium (controlled by Nav1.5 and SCN5A genes), potassium (Kir2.1, KCNJ2), and calcium (Cav1.2, CACNA1C; RyR2, SERCA2) channels. The electrical current across cardiac cell membranes controls contraction, with the sarcoplasmic reticulum in the body of the cell acting as a calcium reservoir. Defective channel proteins, often due to genetic variants, trigger potentially fatal arrhythmias.

Clinical monitoring with ECGs may detect abnormal heart rhythms, although in some individuals the arrhythmia is intermittent. Long QT syndrome (LQTS) affects approximately 0.4 in 1,000 live births, with morbidity and mortality risks increased by certain drugs, physical exertion, or stress.[17] Treatment with beta-blockers and, when necessary, device therapy (such as implantable defibrillators) has dramatically improved survival for affected families.

CPVT (catecholaminergic polymorphic ventricular tachycardia, affecting about 0.015 per 1,000 live births) and SQTS (short QT syndrome, 0.01–0.03 per 1,000 live births) can be triggered by the release of catecholamines—such as during exercise, emotional stress, or REM sleep—sometimes causing fatal arrhythmias in up to 50% of untreated or poorly managed individuals before age 30.[18]

Epidemiological research by Professor Peter Schwartz (Director of the Cardiovascular Genetics Laboratory at the IRCCS Istituto Auxologico Italiano, Milan, Italy), and colleagues, established links between QT interval prolongation and SIDS. The best available evidence now suggests that about 1 in 10 SIDS cases is associated with LQTS, suggesting that routine infant ECGs could identify candidates for preventive intervention. [19]

Catecholaminergic arrhythmias reveal the enriched risk present even in families without structural heart disease—highlighted by sudden unexplained deaths in children, as well as ventricular fibrillation, and lethal syncope during exertion or stress.[20]

Genomic Sequencing and Pathology

Routine heel-prick (Guthrie) testing for newborns has expanded to encompass soon-to-be routine whole-exome and genome sequencing, leading to highly individualized diagnostic and preventive medicine.[21] Research projects (BabySeq, NSIGHT, Cambridge Sick Child studies) have revealed actionable findings in as many as 21% of cases, impacting not only acute management but lifelong reproductive planning.

These evolving technologies offer hope for more accurate classification of variants (from benign, likely benign, VUS, pathogenic), earlier interventions, and possibly corrective therapy with gene editing or embryo selection.[22]

Summary

Scientific advances in anatomy, genetics, and cardiology have upended previous dogmas about unexplained child death, revealing powerful roles for inherited defects, episodic epilepsy, and cardiac electrical dysfunction. Targeted electrophysiological and genomic screening now provides unprecedented opportunities for individualized risk assessment, intervention, and prevention.

The wide variability in presentation, response, and outcome for these conditions continues to challenge researchers and clinicians, but the progress made over the last half-century stands as testament to the value of persistent scientific inquiry and international collaboration. The legacy is not only a dramatic reduction in unexplained deaths, but a new era of precision medicine guiding the future of child health

CHAPTER 7 –

EXONERATIONS

"Injustice anywhere is a threat to justice everywhere."
— Martin Luther King Jr.

Introduction

Wrongful convictions in unexplained infant and early child deaths, especially those involving allegations of abuse or shaken baby syndrome (SBS), continue to devastate families and damage the credibility of both legal and medical systems. The following cases illustrate not only the forensic and prosecutorial errors that drive these tragedies but also the lived reality of the innocent men and women whose lives—and families—were destroyed.

The Registry of Exonerations: Overview

As of 2026, there are 84 documented cases in the author's private register of exonerations for the death of a baby or child up to two years old, spanning 70 US, 5 UK, 5 Canadian, 2 Australian, 1 Japanese, and 1 Swedish cases. Details of the 84 cases can be obtained from the author.

Of these, 33 were women (mean at conviction age 28 years), and 52 were men (mean age at conviction 25 years), and 28% of females and 13% of males were non-parental carers. The children who died were nearly evenly split between sexes, and the mean age at death was 32 weeks.

Two women and five men were sentenced to death (exonerated after a mean of 16 years in prison); 15 women and 11 men were given life sentences (average 9 years served); the remaining exonerated defendants served an average of 14 years. The mean time in prison before exoneration for the whole 84 cases was 8 years.

Prior pathology included:

- Recent falls: 21 cases

- Prior cerebral pathology (seizure/intracranial bleeding): 9

- Blood disorders: 4 (3 with clotting disorder, 1 with sickle cell anemia)

- Severe vitamin deficiencies: 2

- Congenital renal sepsis: 1

- Sepsis: 2

- Metabolic disorder: multiple cases where later genetic or metabolic testing revealed underlying disease

nduct.

Case Histories

Vincent Benavides

For 25 years, Vincent Benavides—a Mexican immigrant with intellectual disabilities—languished on death row in California's San Quentin prison.[3] In 1991, while caring for his girlfriend's toddler daughter, Consuelo, he discovered her unresponsive and bleeding near a carport. The hospital documented severe internal injuries: duodenum, pancreas, and bowel split, and multiple failed attempts to catheterize her.[3] At autopsy, a pathologist with a history of professional misconduct claimed her abdominal injuries were caused by vaginal and anal rape— contradicted by anatomical knowledge and the witnessing staff, who had induced the trauma in the pelvic region during repeated attempts to catheterize her.

Vincent was convicted on evidence that also included coerced and, at times, recanted witness testimony and lab findings (dirt and grass on Consuelo's clothes and shoes) ignored by the court.[3] Despite compelling new expert opinions and multiple medical recantations, Vincent remained in prison for over a decade after the evidence—both medical and circumstantial—clearly exonerated him.[3] Upon release in 2018, he had lost nearly half his life, with compensation unable to remediate years spent wrongly accused and imprisoned.[1] [2]

Sabrina Butler

Sabrina Butler—a vulnerable nineteen-year-old African American single mother—was the first woman in US history exonerated from death row.[3] Raised in poverty by a single alcoholic mother, Sabrina's adolescence was marked by struggle and resilience.

In 1989, Sabrina returned home from a short run to discover her infant, Walter, not breathing; she frantically sought help, but Walter could not be revived. In shock, Sabrina endured hours of accusatory, unrecorded police questioning and was manipulated into signing a false confession.

At trial, her lawyer—unfamiliar with criminal law and heavily intoxicated—called no expert witnesses. Critical medical records that showed that Walter had a rare genetic kidney disease, and that his injuries were consistent with desperate resuscitation, were not presented or disclosed to the pathologist. In addition, the prosecutor sequestered Sabrina's neighbors so that they could not give evidence on her behalf and breached professional conduct rules by socializing with jurors during the trial.

Convicted by an all-white, mostly male jury, Sabrina spent years in solitary confinement awaiting execution, separated from her surviving son and family.

The death penalty abolitionist, **Clive Stafford-Smith,** took up her cause. After five-and-a-half years of advocacy and a new trial, she was acquitted.

Following her release she received $329,000 compensation. Her experiences were detailed in the book *Exonerated: The Sabrina Butler Story* published in 2012. Showing remarkable resilience, she became a passionate advocate for the abolition of the death penalty and reform of the justice system. She is currently the Assistant Director of Membership and Training for 'Witness to Innocence.' She is married, goes under the name of Butler-Smith, and lives with her husband and two children in Memphis, Tennessee.[4]

Alan Yurko

Alan Yurko's ordeal in Florida became an international cause célèbre. Convicted of murdering his ten-week-old son by shaking, Alan was sentenced to life in prison mainly on the testimony of Dr Shashi Gore, a state medical examiner.[5]

Later, it emerged that Dr Gore had made numerous basic errors: reporting that the baby was black-skinned when he was white, that his head circumference was 22 cm when it was at least 47.5 cm, describing microscopic heart findings on a heart that had been removed for transplant before the examination, and missing a fatal infectious meningitis.

Despite clear evidence of wrongful conviction and calls from international experts to overturn the verdict, Yurko endured years of incarceration before a judge recognized the forensic failings and ordered his release in 2004.

Yurko's wife, Frances, campaigned fearlessly throughout; the psychological and material damage to their family, and to Alan personally, were incalculable and not undone. Dr Shashi Gore was formally barred from performing autopsies by the Florida Medical Examiners Commission in June 2004.[6]

Patricia Stallings

Patricia Stallings, a Missouri mother, was wrongfully convicted in 1991 of poisoning her infant son Ryan with ethylene glycol (antifreeze) after hospital laboratory tests reported metabolites that clinicians interpreted as classic for antifreeze ingestion.[9,11] When she later gave birth in prison, her second baby, David Jr., developed identical biochemical abnormalities despite having had no contact with her, prompting metabolic specialists to re-examine both children.[7]

Specialist testing showed that David Jr. had methylmalonic acidemia (MMA), a rare inherited metabolic disorder in which organic acids accumulate in the blood and can be misidentified as ethylene glycol or related toxins by certain assays. Re-analysis of Ryan's stored samples revealed the same MMA pattern, proving that both children were affected by a genetic disease rather than poisoning and that the initial toxicology interpretation was fatally flawed.[8]

On the basis of this new evidence, Stallings's conviction was vacated, she was released after about two years in prison, and all charges were eventually dropped. She and her husband then brought civil actions against the hospital, physicians, and testing laboratory, leading to confidential but widely reported multi-million-dollar settlements for wrongful death and wrongful conviction.* The case has since become a paradigm example of how rare metabolic disorders can mimic poisoning and how misinterpreted laboratory results—combined with tunnel vision—can send an innocent parent to prison.[9] [10]

Shaken Baby Syndrome / Abusive Head Trauma convictions

Seventy-three of the 84 exonerations in the cohort involved SBS/AHT-labelled convictions, typically driven more by contested interpretations of imaging and the absence of prior abuse reports than by direct evidence of assault. In these cases, families were often torn apart by poorly substantiated accusations, with confessions or plea deals sometimes shaped by the threat of harsher punishment and by prosecutors or police pressuring witnesses or failing to disclose test results and medical reports. Even when later expert work identified alternative pathology or natural causes, resistance to overturning convictions remained high, and years often elapsed between recognition of error and actual release. In three cases, defendants entered Alford pleas on legal advice, and two others pleaded guilty to lesser charges at appeal rather than risk continued incarceration despite substantial exculpatory evidence. Overall, roughly one-third of SBS/AHT convictions in the cohort involved some form of official misconduct, including five cases of prosecutor misconduct, six involving withholding of exculpatory evidence, two with police officer misconduct, four with forensic analyst misconduct, two with witness tampering, one with child-welfare agency misconduct, and one with perjury by an official. Yet only three exonerations rested on courts explicitly rejecting SBS/AHT itself as an unreliable diagnostic construct and a further three resulted from the failure to consider causes other than SBS/AHT; far more often, relief was granted on the basis of alternative explanations or procedural error, leaving the formal SBS/AHT framework largely intact in other prosecutions.

Table 1: Factors in Conviction and Exoneration

Conviction Factors	Exoneration Triggers
Outdated/controversial forensics (SBS)	New medical/genetic discoveries
Flawed or single-source expert testimony	Recantation/whistleblower testimony
Police or prosecutor misconduct	Multidisciplinary legal-medical review
Coerced/false confessions	Persistent legal and advocacy efforts
Ignored prior pathology or history	Community and international attention

Summary

These stories reveal that wrongful convictions in child death cases are not random aberrations: they are the predictable result of systemic failings in forensic standards, legal process, and institutional accountability.[11] [12]Families suffer irreparable harm. Science and justice both demand humility, vigilance, and reform. Innocence work—documenting, publicizing, and rectifying these tragedies—remains a public obligation as long as such errors persist.[13]

CHAPTER 8 – MISCONDUCT AND FORENSICS

"The law floats in a sea of science. Most of the time, the law barely keeps its head above water."
— Paul C. Giannelli (legal scholar on forensic science)

Introduction

Wrongful convictions in the United States have exposed not only tragic individual injustices but systematic failures involving official misconduct, flawed forensic science, and institutional inertia. Landmark scandals, government reports, and record numbers of exonerations have transformed how evidence is evaluated and catalyzed historic reforms—but many challenges persist. This chapter examines how these systemic failures produce mass miscarriages of justice, explores key case studies, and highlights recent progress in forensic standards and legislation.

Exonerations and Racial Disparities— Latest Data

By the end of 2023, the National Registry of Exonerations recorded over 3,478 exonerations since 1989, with 153 new exonerations in 2023 and over 2,200 years lost to wrongful incarceration in that year alone.[1] Black Americans remain grossly overrepresented among those exonerated (53–60%)—despite making up about 13–14% of the general population.[2] Whites made up about 32%, Hispanics 12%, others around 3%.[3]

Most falsely convicted were accused of serious offenses: murder (38%), drug crimes (18%), sexual assault (11%), child sexual abuse (9%), robbery (5%), and other crimes (20%).

Primary contributing factors in 2023 were[4]:

- Official misconduct (77%)
- Perjury or false accusation (64%)
- Mistaken witness identification (33%)
- Inadequate defense (33%)
- False or misleading forensic evidence (29%)
- False confession (15%)

Mass Exonerations: Major Cases

Annie Dookhan Drug Lab Scandal (Massachusetts, 2013–21):

A chemist at the Hinton drug lab, Dookhan falsified and fabricated results on an industrial scale. Investigations revealed misconduct in tens of thousands of tests; over 21,000 convictions were vacated and dismissed, the largest incidence of exoneration in US history.[5] [6]

Sonja Faak Case (Amherst, MA, 2014–18):

Farak tampered with, stole, and used evidence at work, contaminating and falsifying thousands of samples. More than 13,000 cases were ultimately overturned.[7]

Baltimore Gun Trace Task Force Scandal (2017):

Eight officers were convicted of wide-ranging corruption—planting evidence, false arrests, theft. More **than** 800 convictions were vacated as a result.[8]

Danziger Bridge Cover-up (New Orleans, 2005-2013):

New Orleans Police Department. shot unarmed civilians post-Katrina, then planted evidence and lied in reports. Multiple officers were convicted after a sweeping federal review.[9]

The Forensic Science Crisis—Report and Reform

The 2009 National Academy of Sciences report was a critical turning point. It found that many forensic disciplines (fingerprints, hair, fiber, bloodstain pattern, bite marks, tool marks, etc.) lacked empirical scientific validation, clear error rates, standardized methodology, or robust accreditation.[10]

Major findings and ongoing issues[11]:[12][13]

- Examiner bias and subjectivity remain in pattern/impression evidence.

- Population-level statistical databases and error rates are missing in many fields.

- Training, accreditation, and publication standards remain inconsistent nationwide, especially in under-resourced states.

- DNA analysis stands out as the most reliable technique, but even this can be compromised by contamination, chain-of-custody errors, or poor laboratory practice.

Recent Reforms (2020–2023):

- The NIST and the DOJ's Forensic Science Committees have promoted new standards, including proficiency testing and accreditation, though not all states have adopted these reforms.

- Recent federal grant programs have increased funding for forensic lab research and modernization, including NIJ and BJA initiatives to expand DNA capacity and public-lab research.[14][15]

- Separate federal proposals have sought to fund broader recording of custodial interrogations and to strengthen infrastructure for managing and disclosing forensic-evidence quality issues, though many of these reforms remain at the bill or policy-proposal stage. Separate federal proposals have sought to fund broader recording of custodial interrogations and to strengthen infrastructure for managing and disclosing forensic-evidence quality issues, though many of these reforms remain at the bill or policy-proposal stage.[16][17]

- Several states—California, Texas, Illinois, New York—require post-conviction DNA testing, improved evidence access, and regular crime lab audits.[18]

Yet, major gaps remain:

- Less than two-thirds of US crime labs are accredited by an independent agency.

- Individual practitioners can still testify based on experience rather than validated scientific protocols in some jurisdictions.

- Misconduct—including the Dookhan and Farak scandals—shows that transparency, monitoring, and robust disclosure rules are critical for every stage of prosecution.

The Limits (and Power) of DNA

Modern DNA identification—especially using 13+ STRs (the analysis of at least 13 Short Tandem Repeat (STR) loci in a DNA profile)—provides a reliable way to confirm or exclude suspects. But success depends on sound evidence handling, cross-contamination safeguards, and the willingness to test and retest in post-conviction cases. Exoneration casework repeatedly demonstrates that even the best science is only as effective as the systems and honesty underpinning it.[19] [20]

Summary

Exoneration data over the last decade proves that official misconduct and dubious forensics are not rare, but deeply embedded. Mass exonerations in

Massachusetts, Baltimore, and New Orleans highlight systemic vulnerability. DNA technology has revolutionized innocence work but is no panacea when the larger legal system remains flawed.

These lessons set the stage for the next chapter, which details the birth and evolution of innocence projects—specialized organizations whose mission is to test, retest, and fight for the wrongly convicted using the lessons learned from forensic breakdowns and systemic misconduct.

CHAPTER 9 – PROVING INNOCENCE

"Hope is being able to see that there is light despite all of the darkness."
— Desmond Tutu

Introduction

The quest to prove one's innocence after a wrongful conviction remains one of the most challenging—and consequential—struggles in modern criminal justice. While the last thirty years have seen a revolution in awareness, advocacy, and scientific methods for exposing wrongful convictions, the scale of the problem, especially in the United States, remains immense. This chapter examines the evolution and global reach of the innocence movement, the repeated patterns underlying false convictions, and—most importantly—for those still incarcerated or fighting from the outside, what steps can be taken to begin the journey to exoneration.[1]

The Scope of the Problem: Innocence, Exonerations, and the Global Movement

Although the United States is home to the largest and best-known innocence projects, wrongful convictions and their remedy are a global issue. From the Boorn brothers wrongly convicted in Vermont in 1819 and exonerated after the missing man was found alive[2] to the most recent cases documented each month, it is clear the possibility of injustice is never far from even "advanced" legal systems.

Over the last 30 years:

- Over 3,700 exonerations have been formally documented in the US since 1989, but this number is believed to represent only a small percentage of the true total.[3]

- Reliable studies now estimate 20,000 to 100,000 people—1 to 5 percent of the prison population—may be innocent but remain behind bars.[4] [5]

- Hundreds—likely over a thousand—are believed to be wrongly imprisoned for child deaths, including convictions based on now-controversial diagnoses such as "shaken baby syndrome" (SBS) and abusive head trauma.[6] [7]

- **Canada:** Innocence Canada, founded in 1993, has overturned dozens of convictions, sparking reforms. [8]

- **UK:** The Innocence Network UK, despite hurdles, is active at over 30 law schools and has fostered review and public awareness. [9]

- **Australia, Ireland, Germany, Sweden, Japan, Mexico, and many more:** All have active innocence efforts, some within universities, others with journalistic or government support. [10]

- **Latin America and Spain:** The Red Inocente alliance works to expand post-conviction rights across legal systems often hostile to review. [11]

- The United Nations and international legal bodies increasingly recognize wrongful conviction as a matter of basic human rights, with ongoing campaigns for post-conviction review in all countries and for standardizing compensation for exonerees. [12] [13]

Timeline – The Global Innocence Movement

1819

First recorded US exoneration: Jesse and Stephen

Boorn (Vermont), released after the "victim" was found alive.

1983

Jim McCloskey, after the exoneration of Jorge De Los Santos (New Jersey), founds Centurion—the world's first dedicated innocence organization.[14]

1986–87

Barry Scheck and Peter Neufeld help exonerate Marion Coakley in New York, setting the stage for DNA-based challenges.[15]

1992

Scheck and Neufeld launch the Innocence Project at Cardozo School of Law, NYC.[16]

1993

Canada's Association in Defence of the Wrongly Convicted (now Innocence Canada) is founded.[17]

1999–2000

DNA science begins revolutionizing post-conviction review; Innocence Project records its first US death row exoneration.[18]

2002

Griffith University Innocence Project starts in Australia.[9]

2004

Innocence Network UK is established, bringing organized innocence review to British law schools.[8]

2009

Irish Innocence Project starts at Griffith College, Dublin.

2012

National Registry of Exonerations (USA) launches, tracking all known US exonerations since 1989.[19]

2013–Present

Red Inocente and similar networks expand into Latin America and Spain, with new projects developing across Europe, Asia, and Africa.

2024

Innocence Network includes over 70 member organizations worldwide; over 3,700 exonerations documented in the US since 1989, with 147 in 2024 alone.[20]

How the Innocence Movement Works—and the Lessons for Families

Innocence projects are not all alike, but their core approaches share common patterns:

1. **Intake and Initial Assessment**

   - Screening: Most innocence projects screen hundreds of requests per year. They focus on claims with strong evidence of actual innocence (not just procedural errors).

- Documentation: Prisoners, relatives, or concerned supporters submit case records, trial transcripts, appellate rulings, and a summary of the claim.

- Eligibility: Priority often goes to cases involving DNA or other new scientific evidence; however, some projects will work with non-DNA cases (false confessions, eyewitness errors, junk science).

2. Advice:

- Gather and preserve every possible court record, transcript, and correspondence.

- Maintain a clear timeline of the case.

- If new evidence exists (DNA, alibi witnesses, recanting experts), highlight this in your application.

- Seek outside support for advocacy.

3. Investigation and Advocacy

- Re-investigation of the case, contacting witnesses, seeking forgotten evidence, recruiting experts.

- Collaborative work by lawyers, investigators, scientists, and journalists.

- o Legal strategy may focus on new evidence motions, post-conviction relief, clemency, or pardon applications.

4. **Litigation or Executive Action**

   - o **Habeas corpus** at state/federal levels if original legal process was flawed or new evidence arises.

   - o **Appeals for new trials** based on ineffective assistance, misconduct, or new science.

   - o **Clemency applications** to governors or boards; media and public campaigns are sometimes critical.

   - o **Reality:** Exoneration takes 8–14 years on average—often in isolation, with loss and uncertainty.

What Would It Take to Overturn Tens of Thousands of Convictions?

True justice requires a system-wide overhaul, including:

- **Mandatory post-conviction review** for all convictions.

- **Open records and evidence preservation** policies.

- **Independent review boards** to assess claims outside of political influence.

- **Universal right to post-conviction counsel.**

- **Tighter forensic standards**—refusing courtroom use of discredited science.

- **Large-scale innocence audits** and review commissions.

- **Incentives, not punishment, for officials who admit to error.**

- **Comprehensive reentry support** for those freed.

- **A cultural shift:** Treating actual innocence as a supreme concern of justice, not an administrative afterthought.[18,19]

Summary

Chapter 9 traces the emergence of the modern innocence movement from early, isolated exonerations to a coordinated global effort to confront wrongful convictions. It highlights how advances in DNA testing, forensic science, and organized advocacy have exposed thousands of miscarriages of justice, while research suggests that tens of thousands more innocent people remain imprisoned worldwide—many in child-death cases involving discredited theories such as shaken baby syndrome. The chapter explains how innocence projects operate, from initial screening and reinvestigation to litigation and clemency campaigns, and emphasizes the emotional and temporal cost: exoneration often takes a decade or more, even when compelling new evidence exists. Ultimately, it argues that achieving meaningful justice will require systemic reform—open post-conviction review, stronger forensic standards, independent commissions, guaranteed counsel, and robust re-entry support—alongside a cultural commitment to treating actual innocence as a core priority rather than an inconvenient exception.

CHAPTER 10 – PATHWAY TO JAIL

"Society judges a woman by the children she raises and a man by the children he sires."
— Emma Cunliffe (from "Murder, Medicine and Motherhood")

Introduction

Kathleen Folbigg's pathway to jail was shaped by the decisions and personalities of four men: forensic pathologist Dr. Alan Cala, Detective Bernard Ryan, Crown Prosecutor Mark Tedeschi, and her husband Craig Folbigg. Each brought distinct professional and personal motives to the investigation and trial, ultimately shaping a prosecution built as much on social suspicion and gendered bias as on evidence. This chapter explores their roles and the cascading missteps, conflicts, and vendettas that led to Australia's most notorious miscarriage of justice.[1]

Alan Cala

Dr. Alan Cala became a fellow of the Royal College of Pathologists in 1994 after only six months' formal training, well short of the traditional two years. He began working at the Department of Forensic Medicine, Sydney, in 1995.[2]

In 2001, a Channel 9 documentary exposed the illegal retention and distribution of body parts—including brains—from the Glebe institute, sometimes for research and without family consent. Cala was implicated in conducting an unapproved experiment in which he struck a corpse's head with a hammer, and was the only pathologist who rejected the commission's findings when the department was dissolved.[3]

In 2003, Cala was named Chief Forensic Pathologist in South Australia. His tenure was rocked by further scandal: a televised call for inquiry into his conduct, a notorious case in which he labeled a homicide by stabbing as suicide, and unauthorized retention of a victim's remains after a fatal explosion.[4]

By 2007, Cala was formally reprimanded and fined at the highest level by the NSW Professional Standards Committee for presenting the wrong corpse's findings in a murder case.[5] Political scrutiny forced his resignation in South Australia and he returned to forensic medicine in Newcastle.[5]

Bernard Ryan

Detective Senior Constable Bernard Ryan, a local officer with limited murder investigation experience, led the inquiry into the Folbigg deaths. Ryan pursued the case with exceptional zeal, successfully pressuring Craig Folbigg and his sister Carol into becoming witnesses for the state. Folbigg's conviction became Ryan's springboard to a high-profile career in the NSW Police, eventually being promoted to Superintendent.[6]

Years later, Ryan's personal life was marred by allegations of domestic violence and intimidation in a subsequent relationship. He was suspended but acquitted after conflicting evidence at trial; the magistrate conceded she would have ruled differently by civil, rather than criminal, standards.[7]

Mark Tedeschi

Tedeschi, the Crown Prosecutor, has faced repeated criticism over his aggressive approach, reliance on rumor, and conduct as a courtroom advocate. In at least one overturned conviction, a Court of Appeal judge described Tedeschi as "egoistic and defiant... lacking insight" and strongly suspected him of trying to improperly influence jurors through "smear, innuendo or speculation."[8]

Tedeschi's rigid, adversarial stance was highlighted when he instructed colleagues that Crown prosecutors should never voluntarily concede ground to the defense, earning a public rebuke from the NSW Bar Association for misunderstanding the prosecutor's independent ethical duty.[9] He resigned as Senior Crown Prosecutor soon after a backlash over these remarks.

Craig Folbigg

Kathleen's husband Craig is portrayed as emotionally immature and controlling, rocked both by grief and jealousy. He sought counseling for the children's deaths, but resented Kathleen's apparent resilience. Their marriage—strained by recrimination and unresolved trauma—unraveled completely after Laura's death. Craig's testimony would become erratic, shifting with encouragement from police, and he later admitted to lying under police pressure and out of spite.[10]

Despite knowing new genetic evidence could help clear Kathleen's name, Craig refused to cooperate with exoneration efforts. At inquiries, his counsel argued for her continued guilt, and after Kathleen's pardon, he publicly doubted her innocence.[11] Craig died in March 2024, weeks after her full exoneration, while undergoing cancer treatment.

The Police Investigation

Laura's death was a turning point. Dr. Quang Tuan Au, treating general practitioner, alerted police due to the rarity of four unexplained deaths in one family. Detectives began a sweeping re-examination of all four children's medical records. Dr. Cala reviewed all prior autopsies, and Craig provided Kathleen's diaries to Detective Ryan—fuel for a new theory of maternal homicide.[12]

Inquiry records and media analyses show that Detective Ryan's handling of Craig's statements often emphasized and recorded only those details supportive of the prosecution theory, while contrary or exculpatory elements were not explored or presented with equal weight." Jealousy and grief made Craig susceptible to Ryan's influence; he vacillated repeatedly, admitting in writing that some of his allegations were fabrications born of bitterness and rejection.[13]

Seeking additional expert support, Ryan secured controversial opinions from Australian SIDS campaigner Dr. Susan Beal, and from Dr. Peter Herdson, who dismissed congenital and acquired disease explanations in all cases and declared "intentional suffocation" the likely cause.[14] [15]

Ryan also solicited forensic opinion from overseas experts—including Dr. Janice Ophoven,[16] a vocal advocate of Meadow's Law, and Professor Peter Berry—who advanced theories of homicide based on contested statistical and pathological reasoning.[16] Surveillance and wiretaps followed.

The Diary and Psychological Profiling

Kathleen's diary—lent to police by Craig—became central to the prosecution. Her private writings, while tonally filled with sadness and self-doubt, were portrayed by prosecution experts as admissions of guilt.

Police employed forensic psychologist Rozalinda Garbutt, who never interviewed Kathleen but nevertheless supplied a highly speculative psychological profile. Garbutt was clearly influenced by Meadow's Law as she provided a quote that, "There are some who say one infant death is SIDS, two leaves a big question mark and with three you yell murder."

Garbutt made a number of false claims: she maintained that "Kathleen mentioned to very few people even the fact that she has had children, friends were even unaware that they existed;" She erroneously claimed that "Kathleen discourages the SIDS support groups assistance and is hesitant to accept monitoring from doctors including attending the Westmead Hospital;." she asserted that Kathleen detached herself emotionally from the children when they were sick.

Garbutt concluded that Kathleen was an emotionally detached woman prone to "dark moods," and given her diary entries "homicide is a more plausible explanation."[17]

The Surveillance Tapes

In the lead-up to Kathleen Folbigg's arrest, police covertly installed listening devices within the family home, resulting in approximately 500 hours of recorded audio across around 200 tapes. Despite this extensive surveillance, only a small fraction of the material was ever transcribed. Police identified one notable incident: a loud noise accompanied by Kathleen's outburst, "I should have fucking done what I was gunna do, stuck it underneath that." Investigators interpreted this ambiguous statement as a reference to the diary they had recently uncovered in the home, despite lacking any corroborating evidence. This recording became a focal point, illustrating the prosecution's readiness to draw guilt from scant or circumstantial material.

The tapes also exposed the internal dynamics of the Folbigg household and, crucially, aspects of Craig's personality. During several conversations, Craig directly admitted to acting out of spite and emotional turmoil. He acknowledged to Kathleen that, in the wake of losing both his daughter and his marriage, he was motivated by "hate and spite and anxiety and grief" when giving statements to police. At one point, he candidly described his decision to implicate Kathleen not as a pursuit of justice, but as a form of retribution—stating his intention was to "fix" Kathleen's life in response to his own suffering. He ultimately admitted he could not follow through, conceding that he knew she was not responsible and expressing reluctance for her to be punished for something she had not done.

Craig revealed a further dimension of his character and moral flexibility during another surreptitiously recorded exchange. He explicitly connected his approach to selling cars with a willingness to exploit personal tragedy. He confessed he would use any means necessary—whether relating to his children's lives or deaths—to gain an advantage in his profession, bluntly describing himself as capable of deviant and opportunistic behavior for personal gain.

These recordings demonstrate how Craig's testimony was colored by deep personal grievances and a willingness to manipulate both circumstances and narrative for his own benefit. The police's selective interpretation of the tapes and their focus on incriminating fragments, while neglecting the wider context of Craig's acknowledged deviance, highlight the problematic evidentiary basis on which guilt was ultimately constructed.

Toward Arrest and Committal

By late 2000, despite mixed expert opinions and the lack of direct evidence, Detective Ryan submitted the case to the Director of Public Prosecutions (DPP). The DPP considered the case too weak to prosecute, so Ryan advocated for a coronial inquest. Dissatisfied by the coroner's recommendation to charge only for Laura's death, he pressed for and secured Kathleen's arrest on four murder charges.

On April 20, 2001, Kathleen attended the district court and was refused bail. On April 24, she was taken back to a district court for another bail hearing, but the prosecutor successfully argued against her release, using Ophoven's claim that the odds of four infants in one family dying of SIDs was one in a trillion. On May 18, she presented before a Supreme Court judge who released her on bail of $8000, provided she stay with Tony Lambkin, report to police twice a week, not approach any Crown witness, and not apply for a passport or approach within three kilometers of an airport or international departure point.

Denied bail, Kathleen awaited trial in prison. The media and her own foster mother, who cut all ties in a public letter, turned against her.[18]

Anthony Busuttil, a world-renowned professor of forensic medicine at the Edinburgh University Medical School, analyzed the cases and concluded that while the deaths were tragic:

> *"It certainly cannot be said, indeed beyond reasonable doubt, that these deaths were irrefutably due to imposed or induced airways obstruction, as by suffocation." [19]*

Yet his planned testimony never reached court, and the prosecution pressed for a joint trial on all four deaths, presenting the sequence of losses as damning evidence of guilt even as most other explanations could not be excluded.[20]

Justice Wood, after reviewing the diaries and evidence, ruled against separate trials, agreeing that the five episodes (including Patrick's ALTE) shared "coincidences" but cautioning against reasoning based solely on Meadow's Rule probability. The Court of Appeals affirmed, and Kathleen became the primary suspect in each child's death.[21]

Summary

Kathleen Folbigg's journey to trial and conviction was shaped not only by tragedy and medical ambiguity, but also by the ambitions, biases, and misjudgments of the men who drove the investigation and prosecution. The rush to paint multiple unexplained deaths as evidence of guilt relied on outdated forensic methods, adversarial policing, psychological conjecture, and deep-seated cultural suspicion toward mothers outside the norm. The events described in this chapter lay bare the risks of tunnel vision and flawed expertise—dangers that would be forcibly reevaluated by advancing genetic science and innocence movements in the decades that followed.[22]

CHAPTER 11 – THE TRIAL

"A well-known lawyer, now a judge, once grouped witnesses into three classes: simple liars, damned liars, and experts."

– Nature, 1885

Introduction

Kathleen Folbigg's trial in 2003 became a case study in the flawed intersection of expert testimony, circumstantial evidence, and prosecutorial ambition. Set against the backdrop of unimaginable loss and absent direct proof, the case pivoted on ambiguous diary entries, the contested opinions of "experts," and reasoning that stretched the limits of probability.[1]

Proceedings and Charges

The trial commenced on April 1, 2003, in the Supreme Court of NSW before Justice Barr and a jury of 12. Crown Prosecutor Mark Tedeschi prosecuted; Peter Zahra appeared for the defense. Charges included:

- Count 1: Murder of Caleb (20/2/89)

- Count 2: Maliciously inflict grievous bodily harm on Patrick (18/10/90)

- Count 3: Murder of Patrick (13/2/91)

- Count 4: Murder of Sarah (30/8/93)

- Count 5: Murder of Laura (1/3/99)

Twenty-two expert witnesses testified over 29 days. The prosecution leaned on the speculative behavioral profile of forensic psychologist Rosalinda Garbutt.

Jury Problems and Pretrial Issues

The fairness of the jury was compromised. One juror's acquaintance—a law student who had done some work on Kathleen's case for Legal Aid and had transcribed some of the surveillance tapes recorded in the Folbigg home—offered an unsolicited opinion of Kathleen's guilt. Despite defense objections, the judge kept the juror. Some jurors privately researched family background and death science, discovering Kathleen's father's crime and exploring technical issues like postmortem cooling.[2]

Crucial information was withheld only selective excerpts of 500 hours of police recordings were shared; the rest became available to the defense only decades later.[3]

Building the Case: "Recurrence" and Probability

Tedeschi, with no evidence of direct harm, weapon, or motive, asserted smothering as the mode of death absent any proof of such. The prosecution built its case on the improbability of multiple SIDS or unexplained deaths in a single family, and the flawed assessment of Kathleen's personality dynamics by Ms. Garbutt.

Though barred from referencing Meadow's disproven maxim, Tedeschi had every expert witness claim that such natural recurrences were unheard of, despite existing literature showing dozens of such cases worldwide. Yet not one witness was asked the foundational question, "Have you researched the literature on recurrence in a single family." This lack of context misled the jury about true statistical probability.[4] Tedeschi repeatedly invoked lightning-strike analogies, and the judge reinforced the theme of unmatched rarity in his summing up.

The Medical Evidence

Caleb Folbigg:

The defense failed to highlight the medical complexity of Caleb's delivery and neonatal illness, which included high-risk forceps use, respiratory distress, air leak (pneumomediastinum), and suspected laryngomalacia. Instead, the prosecution depicted Caleb as healthy until fatally smothered.[5]

Patrick Folbigg:

The Crown argued that both Patrick's ALTE and later death stemmed from suffocation. However, defense and even some prosecution experts agreed epilepsy was a plausible, even likely, cause—especially given Patrick's rapid and full early recovery, incompatible with hypoxic brain damage from smothering.[6] [7]

Sarah Folbigg:

Craig claimed at trial that Kathleen took Sarah from their bedroom, suffocated her, and put her back. Before trial, he told police this was a lie.

Sarah's autopsy showed airway obstruction and rare uvula displacement, but the trial focused instead on superficial abrasions and anatomical details, overlooking clinical and literature evidence that suggested an innocent origin.

Laura Folbigg:

Laura's monitoring revealed serious heart and breathing irregularities, yet this was ignored or actively excluded in trial evidence. Events observed by babysitters—episodes of collapse—were not emphasized for the jury.

All Four Children

Tedeschi told the jury:

"They were all discovered dead or moribund by their mother at, or shortly after death when they were still warm to the touch, and two of them still had a heartbeat. So, they were found by her very shortly, literally minutes, after the cessation of breathing."

In reality, both Caleb and Laura were documented as cold when first examined, and warmth can remain for hours postmortem, depending on the environment, a fact that any competent prosecutor would have known. Of the two allegedly found with a heartbeat, Patrick had been resuscitated after his ALTE; Laura's was an agonal rhythm—a failing heart's final, ineffectual beats.

Dr Alan Cala's Evidence

Cala maintained in court that Laura's myocarditis was only of mild degree and could not have contributed to her death. But it later transpired that after the autopsy he showed slides of Laura's heart to Professor Johan Duflau and others, seeking their advice, and all opined that the degree or myocarditis was sufficient to have caused her death. Cala later claimed to have no recall of this meeting.

Character, Diaries, and "Evidence of Demeanour"

Without evidence of abuse, Tedeschi characterized Kathleen as cold and indifferent, turning to ambiguous diary entries and comments on emotional "flatness." Defense witnesses generally described her as doting, traumatized, and conventional in her maternal devotion. Cross examination revealed that even negative witnesses had previously described her as a loving, attentive mother.

Craig Folbigg's testimony changed repeatedly, and he admitted on the stand to lying to police out of anger or emotional distress. Detective Ryan told him "loving, caring mothers" could be killers, an idea echoed throughout the police and prosecution case.

The Diaries

No forensic linguist or grief psychologist was called to unpack the diaries. The prosecution and judge told the jury to view them as veiled confessions; the defense argued they reflected corrosive, misplaced self-blame. The context that Craig often read or knew of the diaries and sometimes wrote in them was ignored. Modern expert review recognizes these entries as trauma artifacts, not evidence of murder.

The Verdict and Aftermath

After less than nine hours, the jury convicted Kathleen of three murders (Patrick, Sarah, Laura), manslaughter (Caleb), and malicious injury (Patrick). Craig, now engaged to another woman, publicly thanked prosecution and jurors for "setting four beautiful souls free to rest in peace." Detective Ryan said the "tragedy... has been finally told."[8] [9]

Kathleen received a 40-year sentence. Multiple appeals were denied, with courts labeling the diaries "chilling, cogent" evidence.[10] [11] Only the passage of two decades, scientific discovery, and determined advocacy would eventually uncover the flaws at the root of her trial.[12]

Summary

The 2003 Folbigg trial was built on selective expert consensus, misconstrued diary entries, and an argument of improbability. Scientific, medical, and legal evidence that might have changed the verdict was minimized or misunderstood. The pervasive error of treating tragedy as proof of guilt would only unravel through hard-won discoveries in science and shifts in public understanding of forensics and justice.

CHAPTER 12 – THE PUBLIC RESPONSE

"When the whole world is silent, even one voice becomes powerful."
— Malala Yousafzai

Introduction

The campaign to free Kathleen Folbigg stands as a testament to the power of public advocacy, interdisciplinary expertise, and persistent questioning of accepted truths. In the years following her conviction, a diverse group of legal scholars, scientists, journalists, and ordinary citizens galvanized support for her cause, challenging the medical assumptions, legal reasoning, and societal biases that shaped her case.

This chapter traces the evolution of that movement—from initial calls for review, through scholarly critique and grassroots mobilization, to the large-scale legal and scientific collaboration that would eventually help overturn one of Australia's most controversial convictions. By examining the individuals and alliances that fueled the push for justice, we see how collective action, rigorous research, and changing public attitudes can spark profound legal and cultural change.

The Public Campaign

Kathleen's plight was examined by academic lawyer Dr. Emma Cunliffe, a fellow at the University of New South Wales and author of *Murder, Medicine and Motherhood*. Cunliffe argued that Kathleen's conviction stemmed from the complex interplay of medical research, expert testimony, media framing, and, crucially, deep-seated social assumptions about 'good mothering.'[1] When a mother is suspected of harm, she wrote, even mundane maternal behaviors become suspect, leading to a constriction of what counts as "acceptable." In her 2011 book, Cunliffe contextualized Kathleen's case within a broader wave of wrongful convictions related to infant deaths.[2] Cunliffe soon became actively involved in efforts to exonerate Kathleen.

Helen Cummings, daughter of Australia's first female mayor, became a prominent campaigner after decades working at the Newcastle Family Court. She published the memoir *Blood Vows* in 2011, recounting her own experiences with violence and advocating for reforms to family law.[3] Cummings recognized that Kathleen's case was fundamentally a trial about motherhood, not murder, and began supporting her, initiating correspondence and visiting her in prison. In 2013, she wrote to NSW Attorney-General Greg Smith, suggesting that the case be reviewed. She was told the options were to apply to the Supreme Court, the Court of Criminal Appeal, or petition the NSW Governor.

Media interest grew. In February 2013, the *Sydney Morning Herald* published Eamonn Duff's article, "New Science Would Let Folbigg Go Free." Pathologist Professor Stephen Cordner told Duff, "It can only be a matter of time before there's a formal review," noting that no evidence supported homicide as the cause of the deaths. Professor Gary Edmond (UNSW) described the 2003 trial as tainted by misleading, outdated evidence, and Cunliffe, referencing the Waneta Hoyt case in the US, said Folbigg was prosecuted "at a moment in time when there was a particularly punitive account of multiple infant deaths in a given family." Professor John Hilton, who performed the autopsies, agreed that a review was warranted—but it would take another decade before authorities acted.[4]

The following day, Joanne McCarthy's "Crusade to free Kathleen Folbigg" appeared in the *Newcastle Herald*. Barrister Isabel Reed, intrigued, reached out to Cummings, who that night delivered a copy of *Murder, Medicine and Motherhood*. Reed and her colleague Nicholas Moir then drew in law students from Newcastle University, inspired by US innocence projects.

Craig Folbigg's family was contacted by the *Inverell Times*, and his brother John responded publicly that everyone deserved a fair trial, but if the case was to be reviewed, "don't waste taxpayers' money"—challenging advocates to "show us what happened to those babies."[5]

Around July 2013, Shaun McCarthy, director of the Newcastle University Legal Centre, joined Reed and Moir in visiting Kathleen in jail. Soon, fourth-year law students were reviewing the case, joined by Dr. Robert Cavanagh, a barrister with extensive experience in public law and royal commissions.[6][7]

In 2014, McCarthy asked Professor Stephen Cordner to review the pathology evidence. Cordner's 122-page report highlighted that no Folbigg autopsy met minimum standards for SIDS post-mortems and that the conclusions reached at trial were therefore unsafe.[9] Cordner also underscored the lack of traumatic injury in any Folbigg autopsy—evidence strongly against smothering, especially as Laura had teeth, which would make injury likely.[8]

Cordner directly rebutted Dr. Cala's claim that Laura's myocarditis was "mild" and unrelated to her death, supplying micrographs to ten independent forensic pathologists; all agreed the myocarditis could have caused death.

Cordner further noted that since 1992, genetic discoveries had shown that many previously unexplained child deaths could be attributed to underlying mutations—an insight that foreshadowed the eventual breakthrough in Kathleen's case.

On June 16, 2015, the legal team submitted a petition to the NSW Governor, stating that incorrect medical evidence, erroneous assumptions about recurrence, and an unsupported diagnosis of murder made the convictions unsafe. Included were major scientific and legal reports challenging the original findings.

Efforts to prompt government action met with repeated delays, leading Isabel Reed to publicly condemn the stonewalling as "unacceptable, unreasonable, and unconscionable."[9]

Genetics, Media, and Persistence

In June 2017, Rhanee Rego, a University of Newcastle student, joined Dr. Robert Cavanagh's office and soon realized the case against Kathleen was nearly entirely circumstantial. Investigating the science, she quickly located research documenting multiple infant deaths in single families, presenting alternative explanations for the Folbigg deaths.[11] Rego's research paved the way for a new approach to the case, and she became Kathleen's close friend and campaign ally.

As progress stalled, the legal team turned to the media, leading to the August 2018 *Australian Story* episode "From Behind Bars," which brought the case to a national audience and aired interviews with

Professors Cordner, Hill, and Cunliffe, as well as a skeptical Nicholas Cowdery, the former Director of Public Prosecutions.[10]

The broadcast renewed interest from David Wallace, a young Australian with expertise in law, finance, genetics, and immunology. Wallace suggested that advances in genomic sequencing could reveal the real cause of the children's deaths. He contacted Reed and Cavanagh, and within months, the team began assembling genetic evidence.

When Rego finished her studies in 2018, she joined the Newcastle law firm representing Folbigg, dedicating long hours to the case and synthesizing the new scientific material for pending inquiry.

Summary

Over more than a decade, a coalition of lawyers, scientists, journalists, and citizen advocates challenged the assumptions and evidence that had sustained Kathleen Folbigg's conviction. The campaign wound from law reviews and quiet lobbying within government to national media exposure and cutting-edge genetic science, ultimately inspiring the new inquiry that would lead to Folbigg's exoneration. The efforts of Emma Cunliffe, Helen Cummings, Isabel Reed, Shaun McCarthy, Rhanee Rego, and a growing team of allies highlight how public advocacy and evidence-based scholarship can trigger systemic change and redemption.

CHAPTER 13 – THE 2019 INQUIRY

"The adversarial process is designed to discover not the truth, but which adversary can make the best showing."
— Irving Younger (legal scholar)

Introduction

The 2019 Inquiry into the convictions of Kathleen Folbigg marked a pivotal point in Australian legal and scientific history. It was not only a testing ground for newly discovered genetic evidence but also a revealing study of how traditional legal reasoning responds to evolving science, advocacy, and doubt. Launched after two decades of mounting controversy and advocacy, the inquiry brought together legal teams, international geneticists, and leading medical experts in a high-stakes attempt to answer whether the deaths of four children could be explained by rare but real genetic mutations, rather than criminal intent. This chapter traces the preparation, conduct, and outcome of the inquiry, highlighting the clashes between scientific complexity and legal convention—and foreshadowing the next decisive turn in the fight for justice.

Preparation

Public and scientific advocacy finally moved the NSW Attorney-General, Mark Speakman, to announce an inquiry into Kathleen Folbigg's convictions on August 22, 2018, acknowledging that three or more unexplained infant deaths could occur in one family from natural causes. Retired chief judge Reginald Blanch QC was appointed as inquirer, but Speakman's launch statements highlighted the emotional toll on Craig Folbigg, not on the accused—reflecting the enduring suspicion that surrounded Kathleen.[1]

Through connections with legal advocate David Wallace, Professor Carola Vinuesa, a globally renowned medical geneticist and pioneer of genomics in Australia, assembled an interdisciplinary team. Vinuesa and her team, building on prior breakthroughs, prepared to conduct modern genetic analysis using new sequencing from available tissue and Guthrie cards.

Genetics sampling and history-taking by Dr. Todor Arsov revealed a family background for both parents suggestive of possible genetic vulnerabilities— yet Craig refused to cooperate with the analysis, declining to provide a DNA sample.

Whole exome sequencing was completed in late 2018. Reviewing the results together, Vinuesa's team discovered the CALM2-G114R mutation—deemed "probably damaging" by analytic software and notable for its absence in healthy control databases. Kathleen, Sarah and Laura possessed the mutation; further analysis revealed a MYH6 variant and, in Patrick, an IDS gene variant associated with metabolic disease. These findings were shared with the inquiry and opposing "Sydney Team" genetics experts, leading to a debate about the interpretation of pathogenicity, penetrance, and clinical significance.[2] [3]

The Hearing and Scientific Controversy

The inquiry hearings began in March 2019, with counsel Gail Furness leading for the inquiry, Jeremy Morris SC for the defense, and dual teams of geneticists from Sydney (Colley, Kirk, Buckley) and Canberra (Vinuesa, Arsov, Cook). The Sydney Team favored strict ACMG Standards for pathogenicity classification; the Canberra Team opposed this, arguing modern genetics required context and allowance for novel presentations.[5]

International cardiac genetics expert Professor Peter Schwartz was consulted. On June 20, 2019, he linked the CALM2-G114R variant to calmodulinopathies causing sudden death in children—referencing new data on another family whose variant affected the exact residue as the Folbigg mutation, with sudden cardiac arrests and deaths among the children. Schwartz concluded that "the accusation of infanticide might have been premature and not correct."[4]

The Sydney Team, in their July 2019 rebuttal, conceded that the CALM2-G114R variant was likely pathogenic but argued standard factors meant it could not account for the girls' deaths—pointing to rarity, sleep onset, and Kathleen's own apparent good health and ECGs. The Canberra Team's detailed response, endorsed by Professors Schwartz and, Toft Overgaard (Professor of Protein Science, Aalborg University, Denmark) countered each point, emphasizing principles of incomplete penetrance, variable expressivity, the presence of sleep-related deaths in the registry, and the significance of syncope in Kathleen's history. They concluded that cardiac arrhythmia from the CALM2 variant was the likely cause of death for Sarah and Laura, while the boys' deaths could also have a genetic cause.[5] [6]

Inquiry Findings

Judicial Officer Blanch's final report (July 2019) acknowledged that new gene mutations and recurrence of unexplained deaths in families were possible but maintained the deaths were not attributable to natural causes. Relying on his own interpretation of registry data, he argued that CALM2 mutations were rare and that sleep deaths in carriers were "even more extremely rare." He dismissed the genetic explanation, stating the only reasonable finding was that Kathleen Folbigg had smothered her children, reinforcing the importance of "tendency and coincidence" evidence and her purported "obfuscation" in interviews and diaries.[7]

Blanch did, however, concede that the 2003 jury had been incorrectly advised about the statistical and medical impossibility of multiple unexplained deaths in one family—a significant, yet fatal, error in reasoning. Nonetheless, because no mutation was found that accounted for all four deaths, he upheld the convictions.

The Eight Blind Spots in the Folbigg Prosecution

Blind Spot	What Experts Stated	The Reality
Overreliance on Circumstantial Evidence	Rare events were presented as proof of guilt, not uncertainty	Discounted alternative scientific explanations
Ignoring Emerging Genetic Science	Genetic and cardiac causes were downplayed or dismissed	Courts failed to consult up-to-date specialists
Misinterpretation of Diaries	Diaries presented as "confessions"	Trauma and grief response misread as guilt
Outdated Forensic Assumptions	"Smothering leaves no trace"	Modern data show most suffocation

Blind Spot	What Experts Stated	The Reality
	accepted uncritically	leaves evidence
Confirmation Bias and Tunnel Vision	Focused early on maternal guilt, narrowing investigation	Alternative hypotheses were neglected
Failure to Separate Law and Science	Science treated as adversarial argument	Needed neutral, peer-reviewed expert input
Inadequate Post-Conviction Review	Initial review and appeals slow or dismissed new evidence	Justice lagged behind evolving science
Assumed Absence of Evidence	Experts claimed "no known cases" of multiple natural deaths	None performed or cited a formal literature search

Each of these blind spots prolonged injustice in the Folbigg case. Together, they make a powerful argument for law–science integration in all criminal investigations.

Aftermath and Scientific Reflection

The outcome sparked dismay among scientists and advocates. Attorney-General Speakman's statement expressed sympathy for the Folbigg family's pain, but provided scant comfort to Kathleen, who was given no warning of the verdict and was forced to suppress her emotions within custody.

Expert witnesses expressed deep frustration with the adversarial, combative tone of the inquiry. Professors Clancy, Vinuesa, and others described aggressive cross-examination, reluctance to allow full expert testimony, and a perception that the judicial process was ill-equipped to weigh complex scientific uncertainty.[8] Vinuesa and colleagues called for better integration of nuance, peer respect, and scientific reasoning in future legal forums, arguing that more cases would arise at the intersection of advanced genetics and criminal law.[9]

Despite her personal disappointment, legal campaigner Rhanee Rego regrouped and ultimately rededicated herself to innocence advocacy—her persistence forming part of the growing "brains trust" of

Team Folbigg and helping drive the continuing push for justice.

Summary

The 2019 Inquiry into Kathleen Folbigg's convictions forcibly exposed the challenge of integrating evolving scientific evidence into a justice system steeped in adversarial logic and outdated concepts. Although a global consensus grew among geneticists and international scientific bodies supporting the innocence hypothesis, the inquiry's structure and mindset rendered science secondary to juridical conservatism. The lessons drawn from this episode would inform the successful 2022–2023 campaigns that ultimately led to Folbigg's full exoneration.

CHAPTER 14 – THE EUROPACE ARTICLE

"Science knows no country, because knowledge belongs to humanity, and is the torch which illuminates the world."
— Louis Pasteur

Introduction

The publication of the Europace article in 2021, backed by an international coalition of scientists, marked a turning point for scientific and legal recognition of inherited arrhythmia in the Folbigg case. This chapter details the campaign for its publication, the experimental findings, and the implications for each genetic variant contributing to the family's story.

International Scientific Campaign

The day Reginald Blanch KC released his final inquiry report, Professor Carola Vinuesa mobilized Australian Academy of Science leaders and began rallying global support. In late June 2019, she contacted Toft Overgaard in Denmark, asking him to conduct a laboratory assay of the CALM2 variant implicated in the Folbigg deaths. Overgaard responded enthusiastically, mobilizing collaborators from North America and Europe to investigate the variant's functional effect on heart rhythm. Their assays found that CALM2-G114R sharply reduced calmodulin's ability to handle calcium, and severely disrupted both CaV1.2 and RyR2 channels—key in cardiac contraction—supporting a direct link to fatal arrhythmias.[1]

Assembling the Consortium

By August 2019, Vinuesa and Dr. Todor Arsov began drafting a peer-reviewed article to present their findings. Within weeks, 27 eminent co-authors from 16 major centers—including Professors Malene Brohus, Michael Toft Odegaard and Mette Nyegaard from Denmark and Lia Crotti and Peter Schwartz from Italy—joined the effort. Known within the project as the "Brohus paper," the eventual publication would unite leading voices in genetics, neurology, and cardiology from four continents.[2]

During ongoing lab reanalysis, the Canberra Team identified an ultra-rare Bassoon (BSN) gene variant in the boys, sought confirmatory review from Professor Melanie Bahlo (Walter and Eliza Hall Institute, Melbourne, Australia), and commissioned teams in France to analyze other variants. The French data determined that a KCNAB2 variant was likely benign.

Scientific and Legal Hurdles

Controversy dogged publication. Several journals declined the manuscript due to its implicit criticism of the Sydney Team's conservative interpretation of the CALM2 variant. After a publisher finally agreed (following an anonymous letter criticizing the interpretation of the results, and reviewer demands for anonymization of individuals and locations), the article was published in *Europace* on November 17, 2020, with an explicit link in the references to the Folbigg case.[3]

Variant Analysis Summary

Table 2 summarizes the variant distribution (+* represents unaffected carrier of the variant)

- **CALM2-G114R:** Found in the mother (Kathleen), Sarah, and Laura. The functional effect—much reduced calcium binding and ion-channel regulation—was decisively pathogenic. G114R mirrors another deadly calmodulin variant (G114W), which caused sudden death and cardiac arrest in another family.[4][5]

- **BSN (Bassoon):** All four children carried the A335T variant from Craig; Caleb and Patrick also inherited L1898M from Kathleen. While each is rare, their co-occurrence is extraordinary; data from mice show biallelic (changes or mutations in both copies of a particular gene, one from each parent) BSN disruptions cause early lethal epilepsy, suggesting possible pathogenicity for the sons.[6]

- **MYH6:** Present in the mother, Caleb, and Laura. While considered borderline in pathogenicity, this gene is linked with inherited arrhythmias and congenital heart abnormalities.[7][8]

- **IDS:** The X-linked IDS variant was found in the mother, Patrick, Sarah, and Laura; **pathogenic** mainly in Patrick, paralleling known fatal metabolic disorders (Hunter Syndrome).[9]

Scientific Consensus and Rarity

The global scientific consensus, supported by functional and genetic data, concluded that Sarah and Laura almost certainly died due to a calmodulinopathy from the G114R variant; Patrick may have suffered fatal effects from the IDS variant; Caleb and Patrick's combination of BSN mutations presented a credible mechanism for lethal neurological syndromes. The statistically improbable but not impossible co-occurrence of such ultra-rare variants in the same family solidified the case for exoneration.

Summary

With this landmark publication, the burden of proof fully shifted from Kathleen to the state: the best available science supported inherited arrhythmia and genetic disease as explanations for the children's deaths. The article forced the legal system to confront these realities and catalyzed further advocacy, directly influencing the final, successful innocence campaign

CALM2 G114R	BSN A335T	BSN L1898M	MYH6	IDS	NOTE/SYNDROME
+		+	+	+*	LQTS, IVF, Carrier IDS
	+	+	+		BSN compound het., MYH6
	+	+		+	BSN compound het., IDS hemizygote
+	+			+*	CALM2 pathogenic, Carrier
+	+		+	+*	CALM2 pathogenic, MYH6, Carrier IDS

CHAPTER 15 – THE PETITION

"First they ignore you, then they ridicule you, then they fight you, and then you win."
— Mahatma Gandhi (attributed)

Introduction

The campaign to free Kathleen Folbigg reached a historic turning point in March 2021, when more than 90 eminent scientists—including Nobel laureates and global leaders in genetics and cardiology—signed a petition urging her immediate pardon. This international mobilization marked a dramatic escalation in the struggle between newly uncovered scientific truths and the inertia of legal institutions.

At its heart, the petition directly challenged the justice system's reluctance to reconsider convictions in light of compelling genetic evidence and called on New South Wales to remedy what was increasingly seen as a profound miscarriage of justice.

The International Scientific Petition

In March 2021, a landmark petition was submitted to the Governor of New South Wales, Margaret Beazley, urging the immediate pardon of Kathleen Folbigg. Solicitor Rhanee Rego and barrister Robert Cavanagh coordinated the effort, which quickly garnered the signatures of over 90 eminent scientists—and soon more than 150—among them Australian Nobel laureates Elizabeth Blackburn and Peter Doherty, UK heart rhythm pioneer Professor Reza Razavi, and internationally recognized cardiac geneticist Professor Christopher Semsarian.[1] [2]

While a third Nobel Prize recipient supported the petition, he was unable to sign it in time. The petition concluded:

"The executive prerogative of mercy is designed to deal with failures of the justice system such as this one. It is incumbent on the Governor to exercise her power to stop the ongoing miscarriage of justice suffered by Ms. Folbigg. Not to do so is to continue to deny Ms. Folbigg basic human rights and to decrease faith in the New South Wales justice system."

It further warned that ignoring robust medical and scientific evidence in favor of circumstantial interpretation set a dangerous precedent for justice in Australia.

Immediate and Broader Impact

News of the petition had an immediate social impact. For the first time in 18 years, Folbigg was welcomed by fellow inmates who previously shunned her as a "child killer"—the worst possible label in the prison subculture.

In October 2021, journalist Quentin McDermott reported that leading academic experts—including Janine Stephenson (psychiatry, University of Sydney), David Butt (linguistics, Macquarie University), and James Pennebaker (social psychology, University of Texas)—had reviewed Kathleen's diaries. All found no evidence of criminal intent, deception, mental instability, or rage, but rather the language of an anxious, grieving, self-questioning mother.[3]

The Appeal Court's Ruling

Meanwhile, the NSW Court of Appeal considered a review of the 2019 Blanch Inquiry and on 24 March 2021 dismissed the application. The judgment—criticized as naïve by scientific observers—acknowledged a theoretical possibility of innocence based on genetics, but maintained that the Folbigg cases were "outliers," that the girls' deaths differed from previously reported CALM2 cases, and that no common genetic cause explained all four deaths. The appeal judges explicitly privileged inculpatory "inferences" from the diaries over all scientific uncertainty.[4]This decision provoked a near-instantaneous backlash from the scientific community. That afternoon, the Australian Academy of Science released a statement branding the court's approach inadequate, with three of its most senior members publicly calling for legal reform.[5]

> *"Expert advice should always be heard and listened to. It will always trump presumption."*
> *– Professor Ian Chubb, former Chief Scientist of Australia*

> *"There is now an alternative explanation for the deaths of the Folbigg children that does not rely on circumstantial evidence."*
>
> *– Professor John Shine, Academy President*

> *"The science in this particular case is compelling and cannot be ignored. Despite the new knowledge gained from sequencing the human genome almost 20 years ago, we still have some way to go when it comes to both understanding the complexities of genetic disorders and educating the community about these issues."*
>
> *Professor Jozef Gecz (human geneticist, University of Adelaide)*

Professor Carola Vinuesa also issued a public statement, rebutting the claim that the girls' deaths were outliers when viewed against international clinical registries and peer-reviewed data on CALM-related sudden deaths.[6]

Media, Political, and Scientific Pressure

While media reporting was divided—mainstream outlets covered the scientific controversy, while some Murdoch press articles attacked Folbigg's character and the scientific campaign as manipulative—the scientific momentum did not wane.[7]

The Australian Academy of Science's offer to brief Attorney-General Mark Speakman on genetic evidence was declined. Kathleen wrote to Speakman directly, crediting the scientists for restoring her sense of dignity and making a plea for scientific facts to prevail.[8]

In parallel, new global research on calmodulinopathies appeared; a preprint and later peer-reviewed study from the team of Floyd et al. applied AI-powered variant mapping to pediatric cardiac arrests and validated the high pathogenicity of the Folbigg CALM2 variant. They cited the Folbigg case as a striking instance where genetics cast doubt on an infanticide conviction.[9]

Speakman, questioned before parliament in early 2022, reiterated his reluctance to meet with outside scientists, and placed delays in addressing the petition on the frequency of new evidence submitted by Folbigg's legal team.

Meanwhile, in March 2022, Peter Yates (Centre for Personalised Immunology, ANU) personally briefed NSW Premier Dominic Perrottet, highlighting the disconnect between NSW's scientific leadership and its administration of justice.

The Second Inquiry

On 18 May 2022, Attorney-General Speakman announced a second inquiry, led by former Chief Justice Thomas Bathurst—again emphasizing the suffering and re-traumatization of Folbigg's former husband and family, and prioritizing those perspectives in a carefully crafted public message.[10]

Kathleen's lawyer, Rhanee Rego, only learned of the new inquiry moments before the press conference, receiving official word by email as the news broke live.

Letters and Personal Transformation

After the petition, Kathleen found solace in her supporters and resolved, in correspondence with Speakman, to continue fighting for justice for herself and her children if denied mercy.[11]

"With no hard feelings, but much respect,

Ms. Kathleen Folbigg"

Expanding Calmodulin Science

Ongoing research continued to illuminate the complexity and variable severity of calmodulin variants. In February 2022, a Japanese-led international team led by Koichi Kato described a large family with a CALM-N138K mutant and sudden deaths, showing that even severe variants can exhibit a spectrum of clinical consequences—sometimes with only mild symptoms in adulthood. Further analysis showed that the variant was also affecting the potassium channel, but in such a way that it was modifying the effects of the calcium disturbance to produce less severe LQTS.[12]

Summary

Chapter 15 traces the powerful surge of scientific and public advocacy that transformed the Folbigg case from a local Australian controversy into a global referendum on the role of science in law. Despite being rebuffed by the appeals court—whose adherence to circumstantial evidence over expert consensus drew international criticism—the petition galvanized the Academy of Science, journalists, and new clinical research teams. Scientific opinion coalesced around the genetic explanation for the children's deaths, placing mounting pressure on political and legal authorities to act. This groundswell of expert and community support was instrumental in driving the announcement of a second inquiry, setting the stage for not just Kathleen Folbigg's eventual exoneration, but also systemic reform in how science informs justice.

CHAPTER 16 – THE PREPARATION

"Science is not a solitary enterprise. It is a collective effort that transcends borders and boundaries."
— Ahmed Zewail (Nobel Laureate in Chemistry)

Introduction

The lead-up to the Bathurst Inquiry into Kathleen Folbigg's convictions was marked by intensive scientific investigation, international collaboration, and spirited debate. Following the appointment of the Honorable Thomas Bathurst AC QC in May 2022, key scientific and medical parties began assembling and exchanging expert reports, genetic studies, and technical critiques. These preparatory phases were not without discord; opinion among prominent researchers and clinicians sometimes diverged sharply, especially regarding the interpretation of calmodulin variants and assay validity. The inquiry's directions hearings reflected these differences, setting the scene for rigorous, multi-disciplinary scrutiny once proceedings began.

The Appointment and Framework

On May 18, 2022, Her Excellency, Margaret Beazley AC QC, Governor of New South Wales, appointed the Honorable Thomas Frederick Bathurst AC QC to preside over the new inquiry into Kathleen Folbigg's convictions. Bathurst, former Chief Justice and, at the time, Lieutenant-Governor of New South Wales, declared at the outset:

> *"Where a case rests upon circumstantial evidence, the jury cannot return a verdict of guilty unless the circumstances are such as to be inconsistent with any reasonable hypothesis other than the guilt… all the circumstances are to be weighed and considered in deciding whether there is an inference consistent with innocence reasonably open, and… the evidence is not to be looked at in a piecemeal fashion. That is the approach I intend to take in the present case."[1]*

A revolutionary precedent was set: for the first time, the Australian Academy of Science was granted official status as an independent scientific adviser and party to a criminal inquiry. The Academy was empowered to recommend experts, propose technical questions, and guide the inquiry's approach to scientific standards and evidence.[2]

Structuring the Inquiry

Directions hearings set the schedule: the first substantive phase, focused on genetic and cardiac evidence, would begin on November 14, 2022; the second, on February 13, 2023, would address psychological and linguistic evidence related to Kathleen's diaries. Craig Folbigg declined to participate physically or provide DNA, signaling ongoing resistance from within the family.

In the lead-up, tensions flared between the Canberra and Sydney teams. Professors Jonathan Skinner and Edwin Kirk (Sydney) critiqued both the scientific significance and methodology of the Brohus paper, claiming CALM2 could tolerate single amino acid substitutions and that the assays used were not physiologically valid. Professor Toft Overgaard (Aalborg) strongly disagreed, emphasizing that most disease mechanism work in genetics is based precisely on such functional assays, as in the Kato team's calmodulin study.

Rebutting Critiques and New Data

Skinner and Kirk's third claim—that any benign calmodulin variant could appear damaging in these assays (without evidence)—was addressed by Toft

Overgaard. He compared the variant I10T on site 10 of the calmodulin protein (that both the gnomAD and UK Biobank Registry data banks reported was benign) with the pathogenic G114R (Folbigg) and showed that only the latter sharply reduced calcium binding. This analysis arrived just before Overgaard's testimony in Sydney.

Professor Arthur Wilde (Amsterdam), an authority on arrhythmia, questioned whether G114R could cause sudden death given its comparatively lesser effect on calcium currents versus more severe variants like N98S. However, Overgaard, with Professor Wayne Chen (Calgary), provided new RyR2 channel results showing that G114R and G114W in the Folbigg and related registry patients released a third more calcium than controls—greater than prior publications suggested—directly supporting pathogenicity.[3]

New Genetic Findings and Expanded Evidence

In 2018, Professor Sam Chen (Case Western, Cleveland) published work revealing that the severity of LQTS could be influenced by additional genes, with REM2 aggravating calcium overload. In early 2023, Professors Cook and Vinuesa determined all Folbigg children possessed a REM2 G96A variant (absent in

Kathleen), which likely exacerbated the CALM2 effect. This was promptly reported to the Inquiry as highly relevant new evidence.[4]

Around the same time, Professor Line Skotte (Denmark) published an international Brain study (2022) identifying the BSN gene—a variant present in Caleb and Patrick—as a novel susceptibility factor for febrile seizures and early-life epilepsy.[5] This expanded the genetic basis for natural causes, especially for the boys.

International Expert Consensus

In late 2022, Professor Peter Schwartz (Milan) reiterated that pathogenic CALM2 variants can cause death by multiple arrhythmia mechanisms—LQTS, CPVT, IVF, or sudden unexplained death—and that inherited, child-onset and sleep-associated events were now recognized in registry data. Indeed, compared to *de novo* variants, registry cases with inherited variants rose from 33% to 50% between 2019 and 2022. Schwartz sharply criticized narrow readings of the new literature:

> *"The bottom line is that experts formulating conclusions based on 70 cases of CALM mutations to state what is typical and what is not, and especially what 'has never been seen,' should be more careful and humbler. We still*

know so little about Calmodulinopathies! … With 132 cases, the clinical presentation is already changing."

He explicitly declared Laura and Sarah Folbigg's deaths "fully compatible with their CALM mutation." If a CALM mutation is found post-mortem in sudden, unexplained death, "the diagnosis would be 'sudden cardiac death due to a Calmodulin mutation.' End of story."

Schwartz also directly refuted the professional hostility and dismissiveness shown by critics without leading publications in the area:

> *"Theories of the accredited scientist and of the quack are alike to the eyes of the gullible beholders."*
> *– Professor Peter Froggatt (as quoted by Schwartz)*

Summary

Before formal hearings commenced, the Folbigg Inquiry became a focal point for scientific disagreement and critical review. Competing teams from Sydney, Canberra, and abroad presented contrasting analyses of genetic findings, functional assays, and epidemiological data. Heated exchanges—both public and private—preceded consensus on several key points, highlighting both the complexity of the evidence and the passionate investment of international experts. Ultimately, despite initial friction, this preparatory phase ensured that inquiry proceedings were informed by the latest data and diverse perspectives, setting a robust factual and methodological foundation for the historic 2022–2023 inquiry.

CHAPTER 17 – THE 2022 INQUIRY

"Truth will ultimately prevail where there is pains taken to bring it to light."
— George Washington

Introduction

The 2022 Inquiry into Kathleen Folbigg's convictions marked a historic convergence of law and science, assembling an international roster of experts in genetics, cardiology, neurology, psychiatry, linguistics, and forensic pathology. For the first time, the proceedings unfolded in a non-adversarial spirit—with government, scientists, and counsel collaboratively dissecting newly uncovered evidence and evolving scientific consensus. As ground-breaking research, cutting-edge technology, and shifting medical opinions collided in real time, the Inquiry tested the capacity of legal systems to adapt to rapidly advancing knowledge—and set the stage for a final verdict on one of Australia's most controversial criminal cases.

Commencement and Context

The Inquiry began on Monday, 14 November 2022, distinguished by the presence (in person or via video) of leading experts in genetics, cardiac arrhythmias, neurology, psychology, and forensic pathology. Yet it was also marked by drama and constraints: journalist Yoni Bashan reported in *The Weekend Australian* in February 2023 that government underfunding forced private parties to raise AU$80,000 to bring crucial international experts from the UK, Denmark, and Macedonia.[1]

Counsel assisting the Inquiry was Sophie Callan SC, supported by Julia Roy. Both demonstrated expertise in genetics and cardiology, exemplifying the inquiry's new, non-adversarial spirit. In her opening, Callan made clear:

> *"No expert is expected to tell your Honour that the CALM2-G114R variant definitely caused the death of either Sarah or Laura Folbigg. Equally, no expert is expected to tell your Honour that the CALM2 variant could not possibly have caused their death. Rather, the experts divide on where in that middle ground they land in terms of the likelihood that this novel genetic variant could have had any role to play in their deaths."*

Early testimony from Professors Toft Overgaard and Nyegaard, who had just arrived from Denmark, introduced crucial new genetic findings. Hearings were immediately adjourned for three months so other experts could review and comment on these revelations.

Representing Kathleen, Senior Counsel Dr. Gregory Woods outlined her team's position and background. Importantly, the Inquiry unfolded as the field of genetics was advancing rapidly, with sequencing costs plummeting and AI enabling swift analyses—making the gathering and assessment of new data a race against the clock.

Interregnum: Outside Expert Reviews

Before the second stage (February 2023), reports arrived from five additional international geneticists: Professors Arthur Wilde, Dominic Abrams, Pascale Guicheney, Calum MacRae, and Hugh Watkins. Their video-linked testimonies brought a genuine depth to proceedings.

MacRae's critique of the Brohus paper (Jan 2023) dismissed CALM2-G114R as unlikely to cause the deaths, but was rebutted in detail by Professor Nyegaard, who exposed factual errors in his reading of foundational studies. She gently challenged his overstatement of "missense" variants and his misunderstanding that the location of the G114 variant across CALM1/2/3 genes was functionally irrelevant.[2]

[3] Professor Schwartz similarly called out logical fallacies in MacRae's approach: denying arrhythmogenic diversity of CALM variants, discounting CPVT, criticizing the clinical relevance of symptomatic history, and ignoring the import of sleep-related deaths. MacRae's written record was even contrasted with his earlier, contradictory co-authorship that labeled G114R "very likely pathogenic."[4]

Professor Schwartz was also asked to comment on the report of Dr. Cala that criticized the Brohus article's title, language and conclusions. In a 'take no prisoners' onslaught Schwartz opined in his report of January 24, 2023:

> *I was dismayed to realize that Dr. Cala in his entire career has published a total of 24 articles and that his h-index (regarded as possibly the best parameter to assess the impact of scientific publications in the worldwide literature) is 9 (source: Scopus – Jan 23, 2023). The final shock came when I realized that not a single one of his publications did relate to sudden cardiac death or to SIDS.[5]*

Dr. Cala also criticized the report of Professor Peter Fleming, a most respected investigator who opined that there were a number of clinical features that argued against the children being suffocated. Schwartz wrote:

> *"I have read Dr. Cala's criticism and Professor Fleming's perfectly detached responses. I see no reason to comment whenever I happen to observe a battle between a dwarf and a giant. No one in science has depicted these situations better than the late and much respected Professor Peter Froggatt when, commenting on the problems generated in SIDS research by the presence of investigators with quite different levels of scientific expertise, wrote: "The theories of the accredited scientist and of the quack are alike to the eyes of the gullible beholders.""*

Resuming the Hearings

From 13 February 2023, six domains were examined: the possible genetic contributions to the children's deaths; the claim of smothering; Kathleen's cardiac status; Caleb's laryngeal anatomy; Patrick's ALTE; and the infamous diaries.

Genetic Evidence: Sarah and Laura

Twelve geneticists evaluated the CALM2-G114R variant, its impact on calcium, sodium, and potassium channels, and its clinical phenotype (the observable characteristics or traits). New research presented showed G114R, like G114W, can disrupt potassium flow—a finding from a preprint by Kang et al. released just 15 days before testimony.[6]

Of the twelve experts, nine (Arsov, Nyegaard, Toft Overgaard, Vinuesa, Cook, Guicheney, Schwartz, Watkins, Wilde) classified G114R as "likely pathogenic" or "pathogenic." The three dissenters—Kirk, Abrams, and MacRae—accepted it could not be definitively ruled out as a cause but argued it did not satisfy strict ACMG pathogenicity or was statistically improbable across siblings, a line Ms. Callan linked directly to the discredited "Meadow's Law."

Professor Vinuesa outlined the emerging understanding: the same variant can have multiple (and even contradictory) effects; only rarely does phenotype perfectly track genotype; and genetic function is the sum of the primary variant and a vast universe of modifying genes.[7]

Professor Schwartz delivered a compelling summation:

> *"The presence of the calmodulin mutation G114R is sufficient to explain the occurrence of sudden death once other causes have been excluded…*
>
> *The coexistence of a largely asymptomatic mother with infants who die suddenly is not the*

problem. The explanation usually depends on the presence of modifier genes.

…These girls had a double hit, as they had two mutations, not just one. The combination of maternal and paternal DNA has represented a lethal cocktail… This genetic variant is more than sufficient to explain the deaths of the Folbigg girls.

Genetic Evidence: Caleb and Patrick:

Attention was drawn to the biallelic BSN variants carried by both male children, linked to epilepsy and early mortality in animal studies. This was recently buttressed by the work of Professor Line Skotte et al. (Brain, 2022) and Chinese teams, finding enrichment of BSN variants among children with epilepsy and febrile seizures.[8] [9] Early evidence suggested that experimental mice carrying both variants died before birth—a finding echoed by Professor Leanne Dibbens and Nathalie Dehorter's research.[10] [11]

In addition, a newly found CELSR3 variant in all four children (but not Kathleen) and the SEH1L gene variant in Patrick pointed to complex, multigenic vulnerabilities. As Professor Vinuesa cautioned, confirmation of causality would take years of future research.[12]

Smothering: Forensic Pathology Re-examined

Professor Peter Fleming led a consensus, concluding that forensic science did not support the longstanding belief that deliberate smothering leaves no clinical or physical signs. Instead, systematic review of documented cases indicated that most (80%) of proven suffocation deaths in young children showed clear external signs—petechiae, facial injury, or mouth/oral lesions.[13] Professor Cordner, Dr. Cala, and Dr. Orde (Canada) agreed smothering could not be assumed in the absence of positive findings, especially without parent-child violence.

Cardiac Findings and ALTE

Professor Guicheney diagnosed Kathleen as "symptomatic" for cardiac arrhythmias, advocating for preventative beta blockers. Dominic Abrams allowed for mild CPVT but insisted arrhythmia could not be excluded. For Patrick's ALTE, Professor Monique Ryan held that his rapid recovery from his ALTE, lack of multi-organ damage, and negative post-mortem argued against an anoxic event; instead, all findings pointed to a neurogenetic disorder and initial epileptic seizure.[14]

Diaries: Psychiatry, Psycholinguistics, and Grief

For the first time, Kathleen's diaries were forensically appraised by a panel of mental health, psychiatry, and linguistics experts (including Dr. Joanna Garstang, Dr. Betts, Dr. Kerry Eagle, Mr. Patrick Sheehan, Dr. Yumma Dhansay, Dr. Janine Stevenson, Dr. Richard Furst, Dr. Katie Seidler, Dr. Sharmila Betts, David Butt, Professor James Pennebaker, Dr. Kamal Touma, and Professor Anthony Korner). Their overwhelming consensus: the diaries showed no admissions, psychosis, or homicidal patterns, but were consistent with complex bereavement, self-blame, and low self-worth found in mothers who lose children to sudden death.[15]

Submissions and Immediate Judgment

Counsel Assisting (April 2023) submitted that reasonable doubt existed for all convictions. The DPP agreed, acknowledging that the new genetic and forensic evidence was "centrally important" and "beyond the contemplation of science" at the time of trial, and left open the appeal. Even Ms. Folbigg's critics could not refute the science unearthed by the Inquiry.

The second Folbigg Inquiry marked the first time an Australian learned academy- indeed, one of the first in the world—served as an official, independent scientif advisor at this level of legal review. The Australian Academy of Science participation included:

- **Expert nomination and questioning:** Ensuring that the world's foremo authorities on relevant genes, diseases, and assays could be called, a framing evidence directly.
- **Direct peer consultation:** The Academy could submit clarifications a challenges as the evidence evolved, and critiqued interpretations strayi from consensus or scientific method.
- **Methodology guidance:** The Inquiry openly debated, and oft implemented, scientifically accepted protocols, functional assays, a statistical reasoning—reversing the adversarial, dismissive stance of t 2019 process.
- **Global impact:** The case is now widely cited as a test case for integrati expert scientific input after legal appeals are exhausted and as a prompt f systemic law reform.

As a result, not only was scientific evidence put at the center of the process, but t standards of reasoning, transparency, and professional discourse were elevated setting a precedent for future intersections of genetics and justice.

On June 1, 2023, Bathurst issued a preliminary report:

"There was a reasonable doubt as to the guilt of Ms Folbigg for each of the offences… the

coincidence and tendency evidence which was central to the Crown case falls away."[16]

Aftermath: Pardon, Freedom, and Exoneration

On June 5, 2023, Kathleen Folbigg was granted an unconditional pardon and released.[17]

On December 14, 2023, her convictions were quashed, and she was acquitted by the NSW Court of Criminal Appeal.[18]

She declared:

"Today is a victory for science and especially truth… For the last 20 years I have been in prison, I have forever and will always think of my children, grieve for my children, and I miss them and love them terribly."[19]

Postscript

Craig Folbigg, still refusing to accept the science and persisting in his conviction of Kathleen's guilt, died in 2024. Kathleen, with new identity and freedom, began the arduous process of rebuilding the life taken from her—and honoring the memory of her children, finally vindicated by law, science, and time.

Summary

The 2022 Bathurst Inquiry represented a watershed moment in the intersection of law and science. For the first time in Australian history, a criminal inquiry proceeded on a non-adversarial basis, marshalling a global assembly of geneticists, cardiologists, forensic pathologists, psychiatrists, and linguists to re-evaluate the evidence against Kathleen Folbigg. International experts clashed, debated, and ultimately reached consensus: the CALM2-G114R variant carried by both Folbigg girls was a plausible—and likely—cause of their sudden deaths, while biallelic BSN variants in the boys pointed to genetic vulnerability for epilepsy.

The forensic orthodoxy that smothering leaves no trace was overturned.

Kathleen's diaries, so long weaponized against her, were reinterpreted as expressions of maternal grief rather than admissions of guilt.

Counsel Assisting, the DPP, and even former critics conceded the science demanded reasonable doubt. In June 2023, Kathleen was pardoned and released; by December, her convictions were quashed and she was formally acquitted. The inquiry demonstrated that when law opens itself to evolving science, justice—however delayed—can finally be achieved.

CHAPTER 18 –

LESSONS LEARNED

"The future belongs to those who prepare for it today."
— Malcolm X

Introduction

The saga of the Folbigg case does not end with a single exoneration. It is a catalyst—both a warning and an opportunity—for legal reform around the globe. As science advances and societies grapple with the consequences of wrongful convictions, justice systems must evolve to prevent, detect, and correct miscarriages of justice. This chapter surveys the scope of the innocence crisis, practical lessons from recent exonerations, and the major reform trend of 2025: the formal integration of scientific expertise into criminal justice.

Law Reform in 2025

Across the US, UK, Australia, and Canada, legal systems are adopting new models to prevent wrongful convictions in medically complex cases—directly mirroring lessons from the Folbigg exoneration.

Science Advisory Boards:

Inspired by the Australian Academy of Science's historic role in the 2022–2023 Folbigg inquiry, both the United Kingdom and several US states (notably California, New York, and Massachusetts) have now tabled or advanced proposals for official science advisory panels.[1] These panels would assist in criminal appeals involving disputed forensic or medical evidence, ensuring the judiciary has access to impartial, up-to-date expertise.

In the UK: Draft bills (Autumn 2025) propose boards to advise on shaken baby cases, genetic controversies, and SIDS-related convictions.[2]

In the US: State legislatures are considering mandates for independent scientific review before retrials or clemency decisions—particularly when novel biomedical knowledge is involved.[3]

Innocence Review Commissions:

England and Wales are also debating enhanced Innocence Review Boards, modeled after US initiatives in North Carolina and Illinois, to empower broader reexamination of convictions in the light of advancing science.[4]

The Driving Principle:

Legal and scientific voices in major journals and media now emphasize that sidelining domain-specific science is a systemic hazard. There is a strong move toward mandated, neutral science mediation—superseding the adversarial "battle of experts"—particularly in cases of rare disease, genetics, and contested cause of death.

Takeaway

These developments mark a significant, real-time shift. They make clear that justice systems must keep pace with evolving science, reducing the risk of wrongful conviction due to outdated knowledge, and creating more robust mechanisms for reviewing contested expert evidence.

Landmark SBS Ruling

In 2025, the New Jersey Supreme Court considered two "shaken baby syndrome" cases—State v. Nieves and State v. Cifelli—in which fathers were charged with criminal offences after their infant children exhibited the SBS triad of encephalopathy, subdural haematoma, and retinal haemorrhages. During the proceedings, experts explained to the court that the concept of SBS evolved from Professor Ayub Ommaya's 1968 paper and the subsequent publications of Drs Arthur Guthkelch and John Caffey, but that biomechanical scientists had subsequently challenged the theory that whiplash injuries without impact could produce the triad.

In November 2025, the court issued a landmark 6-1 ruling that banned expert testimony on SBS in cases occurring without evidence of an impact injury.[5] The court noted that the Frye standard requires trial courts to determine whether the science underlying proposed expert testimony has gained general acceptance in the particular field to which it belongs. While the SBS concept was accepted by many in the paediatric medical community, it was not currently accepted in the biomechanical engineering community—and biomechanical science was foundational to the diagnosis. Therefore, the State failed to satisfy the Frye standard for admissibility.

The court emphasized that "when the underlying theory integrates multiple scientific disciplines, the proponent must establish cross-disciplinary validation to establish reliability."[11] The ruling represents the first time a US state supreme court has formally barred shaken baby syndrome testimony on scientific reliability grounds, and it has profound implications for similar cases nationwide and internationally.[6]

The Scale of Wrongful Conviction

As discussed previously, perhaps more than 100,000 may be serving time in the US for crimes they did not commit.[7] Yet from the late 1980s to the present, just over 3,500 individual exonerations have been documented.[8]

In other countries with fewer resources or less robust post-conviction review systems, the disparity is starker—far more innocent people remain imprisoned due to procedural barriers and lack of support.

Key Lessons from Exonerations

The road to exoneration is never straightforward. The Folbigg case, and countless others like it, show that actual innocence is not enough to guarantee justice. Success depends on persistence, new evidence, scientific breakthroughs, and organized public advocacy—none of which are assured without sustained effort.

Practical Guidelines for the Innocent and Their Advocates

- Do not give up: Exoneration is difficult and rare, but determined persistence can make the difference.
- Work with experienced legal professionals: Seek advocates with deep expertise in post-conviction innocence law.

- Specialized knowledge is essential for navigating bureaucracy and barriers.
- Contact multiple organizations: Resources are often stretched in innocence work; apply widely to projects and legal clinics to broaden your chances.
- Raise awareness: As the Folbigg case illustrates, ethical public campaigns and media outreach can prompt critical re-examination and draw in expert and community support.

Final Thoughts and Resources

For those enduring wrongful imprisonment or seeking justice after release, the way ahead is fraught but far from hopeless. The global innocence movement continues to grow, reform initiatives are underway on multiple continents, and every exoneration is a lesson for future progress. By staying connected, persistent, and thoroughly informed, you help move justice forward and strengthen the system against future errors. A selection of available resources is given in Appendix III.

Kathleen Folbigg's relentless pursuit of justice provides one powerful lesson. Whatever you do, don't give up.

The Fight to Free Robert Roberson

Robert Roberson is an autistic father who has spent more than twenty years on Texas's death row for the alleged "shaken baby" murder of his chronically ill two-year-old daughter, Nikki. In 2002, Nikki was suffering from undiagnosed severe pneumonia and had been prescribed powerful respiratory-suppressing drugs when she collapsed after a short fall from a bed; her **death** was quickly attributed to shaken baby syndrome, and Roberson's flat affect in the hospital was treated as evidence of guilt rather than a feature of his disability. At his 2003 trial, prosecutors relied on then-orthodox SBS testimony claiming that a specific triad of brain and eye findings could only be caused by violent shaking by the last caregiver, a theory that subsequent research has dismantled.

In the decades since, independent pathologists and other specialists have concluded that Nikki died from overwhelming sepsis, and a bleeding disorder aggravated by inappropriate medication, not abuse, and the former lead detective on the case has publicly recanted and now supports Roberson's innocence. In October 2025, after Texas courts and lawmakers heard extensive new medical evidence and debated the state's "changed-science" statute, the Texas Court of Criminal Appeals stayed Roberson's scheduled execution and sent his case back for reconsideration in light of its own recent shaken-baby ruling. His ordeal illustrates how deeply outdated forensic theories can become embedded in convictions and how, much like the 2022 Folbigg inquiry, only robust, independent scientific review mechanisms can prevent or correct such miscarriages of justice.

APPENDIX I - EVENTS IN THE SBS SAGA

Year	Event/ Development	Key Figures	Citation
1946	First association of subdural hematoma and fractures in infants	John Caffey	1
1962	"Battered Child Syndrome" concept published	Charles Henry Kempe	2
1971	Hypothesis linking infant SDH with shaking—"whiplash injuries"	Arthur Norman Guthkelch	3
1972	"Shaken infant syndrome" causing SDH and fractures	John Caffey	4
1984	Ludwig & Warman define SBS triad	Stephen Ludwig & Matt Warman	5

Year	Event/ Development	Key Figures	Citation
1987	Duhaime et al. biomechanics study undermines shaking as triad cause	Anne-Christine Duhaime	6
1990	Judge Mikva warns of trial bias in US child abuse cases	Abner Mikva	7
1995	"Handle with Care" advocacy leaflet (UK)	Helen Carty & Jane Ratcliffe	8
2001	Geddes et al. publish on non-traumatic causes of triad	Jennian Geddes	9
2001	US National Association of Medical Examiners SBS position paper	NA of Med. Examiners	10
2003	Donohoe review finds inadequate	Mark Donohoe	11

Year	Event/ Development	Key Figures	Citation
	SBS causation evidence		
2003	Geddes III paper finds dural bleeding common in hypoxic deaths	Jennian Geddes et al.	12
2004	Geddes doctrine discredited at appeal	Jennian Geddes	13
2006	Nat. Assoc. of Med. Examiners withdraws SBS position paper	NA of Med. Examiners	14
2007–2010	UK pathologists targeted for SBS skepticism	Waney Squire et al.	15
2009	AAP replaces SBS with "Abusive Head Trauma"	US Academy of Pediatrics	16

Year	Event/ Development	Key Figures	Citation
2012	Oklahoma symposium scrutinizes SBS convictions	Keith Findley	17
2014	Judge Kennelly finds SBS evidence "unscientific", releases Del Prete	Jennifer Del Prete	18
2014	Squier receives GMC hearing notice for SBS evidence	Waney Squire	19
2015	Public Health Group warns against SBS-based prosecutions	Lord Justice Judge	20

APPENDIX II – SIDS RECURRENCE IN FAMILIES

Study	2 Cases	3 Cases	4 Cases	5 Cases	6 Cases
Froggatt et al., 1971[1]	5	1			
Peterson et al., 1980[2]	18				
Jeffrey et al., 1983[3]	10	3			
Rosen, 1983[4]		1			
Diamond, 1986[5]				1	

Study					
Emery, 1986[6]	12	1			
Oren et al., 1987[7]		2	2		
Beal & Blundell, 1988[8]	10	1			
Morild, 1989[9]	6				
Guneroth et al., 1990[10]	5				
Wolkind et al., 1993[11]	26				
CONI, 1998[12]	104	1			
Calenda et al., 2000[13]	11	3	1		
McFarland et al., 2002[14]					1

EagleSmith et al., 2006[15]	**8**	**2**	**1**		
Garstang et al., 2020[16]	**16**	**1**			
Total	**231**	**16**	**4**	**1**	**1**

APENDIX III - ESOURCES

- **Global directory** – Innocence Network (links to all member projects worldwide)
  https://innocencenetwork.org/directoryinnocencenetwork+1
- **United States – Innocence Project** (New York)
  https://innocenceproject.org
  Email: info@innocenceproject.orginnocenceproject+2
- **United States – Pennsylvania Innocence Project**
  https://www.painnocence.org
  Email: info@painnocence.orgpainnocence
- **United States – Mid-Atlantic Innocence Project**
  https://exonerate.org
  Email: info@exonerate.orgexonerate
- **United States – Georgia Innocence Project**
  Website: https://www.georgiainnocenceproject.org
  Email: gip@georgiainnocence.orggeorgiainnocenceproject
- **Australia – Griffith University Innocence Project**
  https://www.griffith.edu.au/arts-education-law/griffith-law-school/learning-teaching/innocence-project
  Email: innocence-project@griffith.edu.augriffith+2
- **United Kingdom – Manchester Innocence Project (example UK clinic)**
  https://www.manchester.ac.uk (search "Manchester Innocence Project" on the site for current contact page)manchester
- **Ireland – Irish Innocence Project at Griffith College Dublin**
  https://www.innocenceproject.ie
  Email: see contact form on website (no single generic email published)innocenceproject

For the most up-to-date list of innocence organizations in your region, visit the Innocence Network directory at https://innocencenetwork.org/directory."wikipedia+1

ABOUT THE AUTHOR

Dr. John Sydney Smith, MD, is a neuropsychiatrist and medico-legal consultant recognized for his work in the evaluation of cognitive and behavioral disorders. He has served as the Director of the Neuropsychiatric Institute at the University of New South Wales and contributed to advancing the field of neuropsychiatry in both clinical and legal contexts. Dr. Smith's expertise includes complex assessments within forensic settings, highlighting his reputation as a leading authority in medico-legal psychiatry.

Throughout his distinguished career, Dr. Smith has engaged in both research and practice, co-authoring publications on brain damage and surgical interventions for epilepsy and severe psychiatric disorders and sharing insights into the interplay between neurological conditions and behavior. He is noted for his commitment to human rights and ethical practice within psychiatry, having participated in significant professional and public discussions on patient care standards.

Dr. Smith's broad influence within the neuropsychiatric and psychiatric professions stems from his leadership roles, scholarly contributions, and dedication to the humane and scientifically grounded advancement of mental health care.

INDEX

REFERENCES

CHAPTER 1

[1] Wikipedia. Kathleen Folbigg. https://en.wikipedia.org/wiki/Kathleen_Folbigg

[2] ABC News Australia. The science that unlocked a rare genetic mutation in the Folbiggs. 4 June 2023. https://www.abc.net.au/news/2023-06-04/kathleen-folbigg-pardon-scientists-dna-unlocked-genetic-mutation/102426796

[3] BBC News. Kathleen Folbigg: Misogyny helped jail her, science freed her. 6 June 2023. https://www.bbc.com/news/world-australia-65810056

[4] ABC News Australia. Kathleen Folbigg's tragic life started long before her babies died. 3 June

2023. https://www.abc.net.au/news/2023-06-03/kathleen-folbigg-life-story-before-wrongful-jailed-pardon/102420716

[5] Bathurst T. Report of the Second Inquiry into the Convictions of Kathleen Megan Folbigg. Sydney: NSW Office of Special Inquiries; 2023. Available from: https://www.nsw.gov.au/nsw-government/projects-and-initiatives/folbigg-inquiry

CHAPTER 2

[1] ABC News Australia. The science that unlocked a rare genetic mutation in the Folbiggs. 2023. https://www.abc.net.au/news/2023-06-04/kathleen-folbigg-pardon-scientists-dna-unlocked-genetic-mutation/102426796

[2] BBC News. Kathleen Folbigg: Misogyny helped jail her, science freed her. 2023. https://www.bbc.com/news/world-australia-65810056

[3] ABC News Australia. Kathleen Folbigg's tragic life started long before her babies died. 2023. https://www.abc.net.au/news/2023-06-03/kathleen-folbigg-life-story-before-wrongful-jailed-pardon/102420716

[4]Murphy DJ, MacKenzie IZ. The comparative risks of rotational delivery with Kielland forceps and vacuum extraction.
Obstet Gynecol. 1998;92(5):777–83. Available from: https://doi.org/10.1016/S0029-7844(98)00258-1

[5] Holinger LD, Sanders AD. Congenital laryngomalacia: Etiology and management.
Ann Otol Rhinol Laryngol. 1991;100(8):623–30. Available from:
https://doi.org/10.1177/000348949110000802

[6] The Conversation. Kathleen Folbigg's children likely died of natural causes, not murder.
2021. https://theconversation.com/kathleen-folbiggs-children-likely-died-of-natural-causes-not-murder-156762

[7] Bathurst T. Report of the Second Inquiry into the Convictions of Kathleen Megan Folbigg.
Sydney: NSW Office of Special Inquiries; 2023. Available from: https://www.nsw.gov.au/nsw-government/projects-and-initiatives/folbigg-inquiry

[8] Blanch TR. Report of the Inquiry into the Convictions of Kathleen Megan Folbigg.
Sydney: NSW Department of Communities and Justice; 2019. Available

fromhttps://www.justice.nsw.gov.au/justicepolicy/D
ocuments/folbigg-inquiry-report-2019.pdf

[9] Colley AF. *Report to the 2019 Inquiry into the
Convictions of Kathleen Megan Folbigg* (26
November 2018). Cited in: 2019 Folbigg Inquiry,
Opening Address of Counsel Assisting, Part 2,
Chapter 4 "Genetics".

[10] ABC News Australia. Kathleen Folbigg's diary
entries show 'grieving mother' not guilty of murder,
inquiry told.
2023. https://www.abc.net.au/news/2023-02-
12/kathleen-folbigg-diary-entries-heard-at-
inquiry/101966798

[11] Blanch TR. Report of the Inquiry into the
Convictions of Kathleen Megan Folbigg.
Sydney: NSW Department of Communities and Jus
tice; 2019. Available from:
https://www.justice.nsw.gov.au/justicepolicy/Docu
ments/folbigg-inquiry-report-2019.pdf

CHAPTER 3

[1] Roser M, Ritchie H, Ortiz-Ospina E. Mortality in
the past: every second child died. Our World in

Data. 2023. https://ourworldindata.org/mortality-in-the-past

[2] World Health Organization. Child mortality and causes of death: Key facts. WHO Fact Sheet. 2022. https://www.who.int/news-room/fact-sheets/detail/levels-and-trends-in-child-mortality-report

[3] Wikipedia. Infant mortality: Historical and global trends. Last updated 2024. https://en.wikipedia.org/wiki/Infant_mortality

[4] Sewall S. Diary of Samuel Sewall, 1686 entry. Massachusetts Historical Society archives. https://www.masshist.org/

[5] Cheever D. Case report presented to the Boston Society for Medical Improvement, March 9, 1863. Harvard Medical School archives. Also cited in: Conolly P. The Water Doctor's Daughters. https://paulineconolly.com/2013/the-water-doctors-daughters-literary-walk/

[6] Conolly P. The Victorian Water Cure Doctors of Great Malvern. Historical review. https://the-malvern-hills.uk

[7] Kopp JH. The Medical Thesis on Thymic Asthma. Heidelberg University Medical Archives. 1830.

[8] Paltauf A. Status Thymo-Lymphaticus Hypothesis and Pathology. Vienna University Faculty Records. 1889.

[9] Wikipedia. Thyroid cancer and irradiation risks. https://en.wikipedia.org/wiki/Thyroid_cancer

[10] Weller RO. The myth of status thymolymphaticus: A historical perspective on sudden death and the thymus.
Am J Dis Child. 1980;134(1):28–33. Available from: https://doi.org/10.1001/archpedi.1980.02130130030009

[11] Medical Research Council (UK). Report on Status Lymphaticus. London: HMSO. 1931.

[12] Editorial. The Lancet on status lymphaticus. The Lancet. 1931.

[13] Weller RO. The myth of status thymolymphaticus: A historical perspective on sudden death and the thymus.
Am J Dis Child. 1980;134(1):28–33. Available from: https://doi.org/10.1001/archpedi.1980.02130130030009

---

[14] Bouchardon P. Celestine Doudet. Paris: Emile-Paul Frères.
1928. https://www.worldcat.org/title/celestine-doudet/oclc/458917079

[15] Conolly P. The Water Doctor's Daughters.
London: Robert Hale. 2013.

[16] Tardieu AA. Étude médico-légale sur les sévices et mauvais traitements exercés sur des enfants. Paris: J.-B. Baillière et Fils; 1860. Available from: https://gallica.bnf.fr/ark:/12148/bpt6k6227928w

1. [17] Templeman C. Deaths from overlaying. Edinburgh Medical Journal.
1893. https://www.ncbi.nlm.nih.gov/pmc/articles/PMC5323076/

2. [18] templeman C. On the sudden death of infants in bed with their parents. *Edinburgh Medical Journal.* 1893. Discussed in: Duncan JR, Byard RW. *A Fresh Look at the History of SIDS.*

3.

[19] The New York Society for the Prevention of Cruelty to Children. Organizational records, historical
summary. https://www.nyspcc.org/about/history/

[20] The New York Society for the Prevention of Cruelty to Children. Organizational records, historical summary. https://www.nyspcc.org/about/history/

[21] Wikipedia. Mary Ellen Wilson's later life and commentary. https://en.wikipedia.org/wiki/Mary_Ell en_Wilson

CHAPTER 4

[1] Roser M, Ritchie H, Ortiz-Ospina E. Child mortality rate over time. Our World in Data. 2025. https://ourworldindata.org/child-mortality

[2] Abramson H. Sudden Death in Infancy in New York City. JAMA. 1944;126(9):585-588.

[3] Woolley E. The role of bedding in sudden infant death. BMJ. 1945;1(4402):722.

[4] Spock B. Baby and Child Care. New York: Pocket Books; various editions, 1946–1958.

[5] National Institute of Child Health and Development. First International Conference on the Causes of Sudden Death in Infancy. University of Washington; 1969.

[6] Choate J. Personal account and advocacy. National SIDS Foundation archives. 1965–1975.

[7] Gilbert R, Salanti G, Harden M, See S. Infant sleeping position and the risk of sudden infant death syndrome: Systematic review. BMJ. 2005;331:326. https://www.bmj.com/content/331/7513/326

[8] Dwyer T, Ponsonby AL. Tasmanian Infant Health Survey: Sleeping position and SIDS. Med J Aust. 1991;154(11):710-713.

[9] Beal S. Paediatrics, SIDS, and sleep positions. Med J Aust. 1986;145(4):211-213.

[10] Willinger M, James LS, Catz C. Defining SIDS and evaluating risk factors: US SIDS prevention campaign outcomes. Pediatrics. 1995;95(5):737-741.

[11] Priyadarshi M. Effect of sleep position in term healthy newborns on SIDS risk. PMC. 2022. https://www.ncbi.nlm.nih.gov/pmc/articles/PMC9326990/

[12] Filiano JJ, Kinney HC. A Triple-Risk Model for the Sudden Infant Death Syndrome. Pediatrics. 1994;100(5):824-830.

13 Steinschneider A. Prolonged apnea and SIDS: Clinical and laboratory observations. Pediatrics. 1972;50(4):646-654. Erratum 1994.

14 Fitzpatrick WJ. Hoyt family case investigation and discovery of serial infanticide. Syracuse Post-Standard. 1994.

15 Southall DP, Samuels MP, et al. Covert video recordings of life-threatening child abuse: Lessons for child protection. Pediatrics. 1997;100(5):735-760. https://pubmed.ncbi.nlm.nih.gov/9326990/

16 Meadow R. Unnatural sudden infant death. Arch Dis Child. 1999;80(1):7-14. https://adc.bmj.com/content/80/1/7

17 Appeal Court Judgments: Cannings, Clark, Anthony. UK Court of Appeal decisions (2003–2005). See reporting in The Independent and BBC News archives.

18 Carpenter RG, Emery JL, et al. Sudden unexplained death in infants: Twins and familial incidence. BMJ. 1984;289(6458):601. https://www.bmj.com/content/289/6458/601

---

[19] EagelSmith RE, et al. Indigenous populations, infant mortality and SIDS recurrence: American Indian health statistics. Pediatrics. 2006;117(4):e758-e765.

[20]UK Court of Appeal Judgment: Standard for SIDS prosecution. Evening Standard. Jan 23, 2004. Reporting on the Angela Cannings case.

CHAPTER 5

[1] Caffey J. Multiple fractures in the long bones of infants suffering from chronic subdural hematoma. Am J Roentgenol Radium Ther Nucl Med. 1946;56(2):163–
173. https://pubmed.ncbi.nlm.nih.gov/20995763/

[2] Kempe CH, Silverman FN, Steele BF, Droegemueller W, Silver HK. The battered-child syndrome. JAMA. 1962;181(1):17–
24. https://pubmed.ncbi.nlm.nih.gov/14455086/

[3] Guthkelch AN. Infantile subdural haematoma and its relationship to whiplash injuries. BMJ. 1971;2(5759):430–
431. https://pubmed.ncbi.nlm.nih.gov/5576003/

[4] Ommaya AK, Faas F, Yarnell P. Whiplash injury and brain damage: An experimental study. JAMA.

1968;204(4):285–
289. https://jamanetwork.com/journals/jama/fullarticle/338923

[5] Caffey J. On the theory and practice of shaking infants: Its potential residual effects of permanent brain damage and mental retardation. Am J Dis Child. 1972;124(2):161–169. https://pubmed.ncbi.nlm.nih.gov/5026202/

[6] Caffey J. The whiplash shaken infant syndrome: Manual shaking by the extremities with whiplash-induced intracranial and intraocular bleedings, linked with residual permanent brain damage and mental retardation. Pediatrics. 1974;54(4):396–403. https://pubmed.ncbi.nlm.nih.gov/4416579/

[7] Ludwig S, Warman M. Shaken baby syndrome: A review of 20 cases. Ann Emerg Med. 1984;13(2):104–107. https://pubmed.ncbi.nlm.nih.gov/6691466/

[8] Duhaime AC, Gennarelli TA, Thibault LE, Bruce DA, Margulies SS, Wiser R. The shaken baby syndrome: A clinical, pathological, and biomechanical study. J Neurosurg. 1987;66(3):409–415. https://pubmed.ncbi.nlm.nih.gov/3819836/

9 Mikva AJ. Opinion in United States v. Bowers, 660 F.2d 527 (D.C. Cir. 1981). Quoted in legal commentary on child abuse trial bias, 1990. https://law.justia.com/cases/federal/appellate-courts/F2/660/527/262869/

10 Department of Health. Handle with care: Preventing shaken baby syndrome. UK Government leaflet. 1995.

11 Plunkett J. Fatal pediatric head injuries caused by short-distance falls. Am J Forensic Med Pathol. 2001;22(1):1–12. https://pubmed.ncbi.nlm.nih.gov/11444655/

12 Geddes JF, Hackshaw AK, Vowles GH, Nickols CD, Whitwell HL. Neuropathology of inflicted head injury in children: I. Patterns of brain damage. Brain. 2001;124(Pt 7):1290–1298. https://pubmed.ncbi.nlm.nih.gov/11408324/

13 Tuerkheimer D. The next innocence project: Shaken baby syndrome and the criminal courts. Wash Univ Law Rev. 2009;87(1):1–58. https://openscholarship.wustl.edu/law_lawreview/vol87/iss1/1/

14 Christian CW, Block R; Committee on Child Abuse and Neglect, American Academy of

Pediatrics. Abusive head trauma in infants and children. Pediatrics. 2009;123(5):1409–1411. https://pubmed.ncbi.nlm.nih.gov/19403498/

[15] Barnes PD, Krasnokutsky M. Imaging of the central nervous system in suspected or alleged nonaccidental injury, including the mimics. Top Magn Reson Imaging. 2007;18(1):53–74. https://pubmed.ncbi.nlm.nih.gov/17607144/

[16] Del Prete J. Exoneration case summary. National Registry of Exonerations. University of California Irvine Newkirk Center for Science & Society; 2014. https://www.law.umich.edu/special/exoneration/Pages/casedetail.aspx?caseid=4377

[17] Findley KA, Barnes PD, Moran DA, Scheck B. Shaken baby syndrome, abusive head trauma, and actual innocence: Getting it right. Hous J Health Law Policy. 2012;12(2):209–312. https://ssrn.com/abstract=2015872

[18] Plunkett J. Fatal pediatric head injuries caused by short-distance falls. Am J Forensic Med Pathol. 2001;22(1):1–12. https://pubmed.ncbi.nlm.nih.gov/11444655/

19 Cory CZ, Jones BM. Can shaking alone cause fatal brain injury? A biomechanical assessment of the Duhaime shaken baby syndrome model. Med Sci Law. 2003;43(4):317–333. https://pubmed.ncbi.nlm.nih.gov/14974414/

20 Leestma JE. Case analysis of brain-injured admittedly shaken infants: 54 cases, 1969–2001. Am J Forensic Med Pathol. 2005;26(3):199–212. https://pubmed.ncbi.nlm.nih.gov/16121071/

CHAPTER 6

1 Centers for Disease Control and Prevention. Infant Mortality Rate by Cause, United States, 2022. National Vital Statistics Reports. Atlanta, GA: US Department of Health and Human Services, CDC; 2023. https://www.cdc.gov/nchs/fastats/infant-health.htm

2 NCBI Bookshelf. Sudden Infant Death Syndrome. StatPearls. Updated July 23, 2023. https://www.ncbi.nlm.nih.gov/books/NBK513346/

3 MedSci. Sudden Infant Death Syndrome (SIDS): State of the Art and Future Directions. Med Sci

Monit.
2024. https://www.medsci.org/v21p0632.htm

[4] Geddes JF, et al. Dural haemorrhage in non-traumatic infant deaths: does it explain the bleeding in 'shaken baby syndrome'? Neuropathol Appl Neurobiol. 2003;29(1):14-22. https://pubmed.ncbi.nlm.nih.gov/12492630/

[5] Gould LJ, et al. Sudden Unexplained Death in Childhood Registry and Research: Surveillance and Findings. Pediatrics. 2024;152(2):e2023067899. https://pediatrics.aappublications.org/

[6] Larcher V, Williams J. Epilepsy as a cause of apparent life-threatening events. Arch Dis Child. 1992;67(8):1029–1033. https://adc.bmj.com/content/67/8/1029

[7] Mendel G. Experiments on Plant Hybridization. Verhandlungen des naturforschenden Vereins in Brünn. 1866. (English translation)

[8] Harper PS. Huntington's Disease. Oxford University Press. 2001.

[9] International Human Genome Sequencing Consortium. Initial sequencing and analysis of the

human genome. Nature. 2001;409(6822):860-921. https://www.nature.com/articles/35057062

[10] Nurk S, et al. The complete sequence of a human genome. Science. 2022;376(6588):44-53. https://www.science.org/doi/10.1126/science.abj6987

[11] Richards S, et al. Standards and guidelines for the interpretation of sequence variants. Genet Med. 2015;17(5):405–424. https://www.nature.com/articles/gim201530

[12] Nyegaard M, et al. A novel calmodulin mutation associated with recurrent cardiac arrest in infants. Eur Heart J. 2012;33(21):2549–2556. https://academic.oup.com/eurheartj/article/33/21/2549/494933

[13] Crotti L, Johnson CN, Graf E, et al. Calmodulin mutations associated with recurrent cardiac arrest in infants. Sci Transl Med. 2013;5(211):211ra160. https://www.science.org/doi/10.1126/scitranslmed.3006980

[14] Crotti L, Spazzolini C, Tester DJ, et al. The International Calmodulinopathy Registry (ICalmR): Relationships among genotype, phenotype, and clinical outcome. Eur Heart J. 2019;40(35):2961–

2971. https://academic.oup.com/eurheartj/article/40/35/2961/5556121

[15] Fukai T, et al. Calmodulinopathy in a Four-Generation Japanese Family. Circ Arrhythm Electrophysiol. 2022;15(3):e010982. https://www.ahajournals.org/doi/10.1161/CIRCEP.121.010982

[16] Jensen HH, Overgaard MT. Calmodulin mutations – lessons for clinical practice. Eur Heart J. Editorial. 2019.

[17] Moss AJ, Schwartz PJ. Long QT Syndrome: From Patient to Gene and Back. Circ Res. 2010;107:462-464. https://www.ahajournals.org/doi/10.1161/CIRCRESAHA.109.211292

[18] Tester DJ, et al. SCN5A, KCNJ2, CACNA1C, and calmodulin in sudden arrhythmogenic death. J Am Coll Cardiol. 2014;64(22):2345-2350. https://www.jacc.org/doi/10.1016/j.jacc.2014.09.021

[19] Schwartz PJ, Stramba-Badiale M, et al. QT interval prolongation and the risk of sudden infant death syndrome. N Engl J Med. 1998;338:1709-

1714. https://www.nejm.org/doi/full/10.1056/nejm199806113382403

[20] Behr ER, et al. Sudden arrhythmic death syndrome: evidence-based diagnostic and treatment approach. Lancet. 2007;369(9563):915-925. https://www.sciencedirect.com/science/article/pii/S0140673607604333

[21] BabySeq Project. Green RC, et al. Clinical Sequencing Exploratory Research Consortium: The BabySeq Project. Genet Med. 2016;18(10):1065-1075. https://www.ncbi.nlm.nih.gov/pmc/articles/PMC5134596/

[22] French CE, et al. Whole-genome sequencing reveals that genetic diagnoses are possible for around 21% of rare disease patients in a national health system. Genome Biol. 2019;20(1):166. https://genomebiology.biomedcentral.com/articles/10.1186/s13059-019-1779-7

CHAPTER 7

[1] Benavides V. Case entry. National Registry of Exonerations. University of California Irvine Newkirk Center for Science & Society; 2018.

https://www.law.umich.edu/special/exoneration/Pages/casedetail.aspx?caseid=5219

[2] Injustice Watch and associated reporting on Vincent Benavides: medical evidence, pathologist recantations, and release from California death row; 2018. https://exonerationregistry.org

[3] Gross SR, O'Brien B, Hu C, Kennedy EH. Rate of false conviction of criminal defendants who are sentenced to death. Proc Natl Acad Sci USA. 2014;111(20):7230–7235. https://www.pnas.org/doi/10.1073/pnas.1306417111

[4] Equal Justice Initiative. Sabrina Butler-Smith: first woman exonerated from death row in the United States. Case biography and commentary; 2015–2023. https://eji.org

[5] National Center for Reason and Justice. The case of Alan Yurko: summary of forensic errors and post-conviction proceedings; 2000–2004. https://ncrj.org/alan-yurko

[6] Florida Medical Examiners Commission. Minutes and disciplinary action regarding Dr Shashi Gore, June 2004. Florida Department of Law Enforcement. https://www.fdle.state.fl.us

[7] Patricia Stallings. Case entry. National Registry of Exonerations. University of California Irvine Newkirk Center for Science & Society; 2020. https://www.law.umich.edu/special/exoneration/Pages/casedetail.aspx?caseid=3820

[8] Settlement with lab in misdiagnosed poisoning. United Press International (UPI) archives. June 8, 1993. https://www.upi.com/Archives/1993/06/08/Settlement-with-lab-in-misdiagnosed-poisoning/6886739512000/

[9] IFLScience. The woman who was wrongfully convicted of murdering her baby and saved by a biochemist: the Patricia Stallings case. 2022. https://www.iflscience.com/the-woman-who-was-wrongfully-convicted-of-murdering-her-baby-and-saved-by-a-biochemist-64986

[10] An inherited metabolic disorder presenting as ethylene glycol poisoning. Case report. Arch Dis Child (or related pediatric journal). 2002. https://pubmed.ncbi.nlm.nih.gov/11953504/

[11] National Registry of Exonerations. Official misconduct and misleading forensic evidence in shaken baby and child-death prosecutions: data

tables and reports; 2017–2024.
https://www.law.umich.edu/special/exoneration/Pages/about.aspx

[12] Innocence Network. Shaken baby syndrome and abusive head trauma cases: overview of exonerations and scientific controversy. Innocence Network report; 2019.
https://innocencenetwork.org

[13] Gross SR, Possley M, Stephens K. The National Registry of Exonerations: lessons from a decade of documenting wrongful convictions. Law & Inequality. 2022;40(2):201–246.
https://lawinequality.org

CHAPTER 8

[1] National Registry of Exonerations. 2023 Annual Report. https://exonerationregistry.org/

[2] Innocence Project. Race and Wrongful Conviction. https://innocenceproject.org/race-and-wrongful-conviction/

[3] National Registry of Exonerations. Report Highlights Racial Disparity. https://michigan.law.umich.edu/news/nati

onal-registry-exonerations-report-highlights-racial-disparity-wrongful-convictions

4 National Registry of Exonerations. Executive Summary 2023. https://exonerationregistry.org/summary

5 NBC News. Epic Drug Lab Scandal Results in More Than 20,000 Convictions Dropped. https://www.nbcnews.com/news/us-news/epic-drug-lab-scandal-results-more-20-000-convictions-dropped-n747891

6 Massachusetts Government. Drug Lab Cases (Annie Dookhan). https://www.mass.gov/info-details/drug-lab-cases-information

7 New England Innocence Project. Amherst Lab Scandal. https://www.newenglandinnocence.org/innocence-blog/2017/6/27/seven-drug-cases-dismissed-in-wake-of-amherst-lab-scandal

8 Gun Trace Task Force Investigation. "Gun Trace Task Force Investigation – Executive Summary and Reports." https://www.gttfinvestigation.org

9 Heritage Foundation. Persistent Forensics Lab Problems Undermine Faith in Our Criminal

Justice. https://www.heritage.org/crime-and-justice/report/persistent-forensics-lab-problems-undermine-faith-our-criminal-justice

10 National Academies of Sciences. Strengthening Forensic Science in the United States: A Path Forward. https://nap.nationalacademies.org/catalog/12589/strengthening-forensic-science-in-the-united-states-a-path-forward

11 National Institute of Standards and Technology. Forensic Science Environmental Scan 2023. https://nvlpubs.nist.gov/nistpubs/waisl/2023/SpecialPublications/NIST.SP.1319.pdf

ASTM International. Forensic Science Standards. https://www.astm.org/standards/forensics/

12 National Institute of Justice. Wrongful Convictions. https://nij.ojp.gov/topics/articles/wrongful-convictions

13 National Association of Criminal Defense Lawyers. Crime Labs and Forensics Reform—Federal Initiatives. https://www.nacdl.org/Landing/Crime-Labs-and-Forensics-Reform

14 NIJ 2023 public-lab research announcementnij.ojpNIJ 2023 public-lab research announcementnij.ojp

15 BJA DNA Funding / CEBR blogbja.ojp

16 Koppl et al., "Creating infrastructure and incentives to increase quality in forensic science"pmc.ncbi.nlm.nih

17 Custodial Interrogation Recording Act (model federal grants bill)congress

18 Innocence Project. DNA Exonerations Lead to Key Policy Changes. https://innocenceproject.org/news/dna-exonerations-lead-to-key-policy-changes-throughout-the-u-s/

19 Boston Bar Journal. From Lab Scandals to Police Scandals https://bostonbar.org/journal/from-lab-scandals-to-police-scandals-lessons-in-resolving-government-misconduct-in-criminal-cases

20 UIC Law Review. Rethinking the Admissibility of DNA Evidence to Prevent Wrongful

Convictions. https://repository.law.uic.edu/cgi/view
content.cgi?article=2942&context=lawreview

CHAPTER 9

[1] National Registry of
Exonerations. https://www.law.umich.edu/special/e
xoneration

[2] Center on Wrongful Convictions, Northwestern
Pritzker School of Law. "First Known Exoneration:
Jesse and Stephen Boorn."

https://www.law.northwestern.edu/legalclinic/wrong
fulconvictions/exonerations/vt/boorn-brothers.html

[3] Innocence Project. How Many Innocent People
are in
Prison? https://www.innocenceproject.org/how-
many-innocent-people-are-in-prison/

[4] Prison Policy Initiative. Mass Incarceration: The
Whole Pie
2025. https://www.prisonpolicy.org/reports/pie2025
.html

[5] PBS Frontline. The Hardest Cases: When
Children Die, Justice Can Be
Elusive. https://www.pbs.org/wgbh/pages/frontline/
the-child-cases/

[6] Brown ST, et al. Shaken Baby Syndrome: as a
controversy in wrongful conviction cases. Albany

Law Rev. 2017–
18. https://www.albanylawreview.org/article/19778-shaken-baby-syndrome-as-a-controversy-in-wrongful-conviction-cases

[7] Innocence
Canada. https://www.innocencecanada.com/

[8] Innocence Network
UK. https://innocencenetwork.org.uk/

[9] Griffith University Innocence
Project. https://www.griffith.edu.au/law/international-centre-for-justice-governance/innocence-project

[10] Norris RJ, Bonventre CL, Redlich AD, Acker JR, Zalman M. Exonerated: A history of the innocence movement. New York: New York University Press; 2020. Available from: https://nyupress.org/9781479817013/exonerated/

[11] Geneva Academy. Toward An International Human Right to Claim Innocence. https://geneva-academy.ch/

[12] United Nations Office on Drugs and Crime. https://www.unodc.org/

[13] Centurion. https://www.centurion.org/

[14] Scheck B, Neufeld P, Dwyer J. Actual Innocence. Doubleday; 2000.

[15] Gannon L, Buitrago S. After Exoneration: The Human Toll of Miscarriages of Justice. Law Soc Rev. 2022.

[16] https://innocenceproject.org/about/

[17] Harvard T.H. Chan School of Public Health. A Matter of Conviction. https://www.hsph.harvard.edu/news/features/a-matter-of-conviction/

[18] Davies RP. How Many People Are Wrongfully Convicted in US Prisons? https://richardpdavieslaw.com/how-many-people-are-wrongfully-convicted-in-us-prisons/

[19] https://www.law.umich.edu/special/exoneration/Pages/learnmore.aspx

[20] https://exonerationregistry.org/sites/exonerationregistry.org/files/documents/2024_Annual_Report.pdf

CHAPTER 10

[1] Inside Story. Judging Kathleen Folbigg. https://insidestory.org.au/judging-kathleen-folbigg/

[2] Exoneration Inquiry: Professional History of Forensic Witnesses. [Folbigg Inquiry site]

³ Channel 9 Sunday. The Body Snatchers. 2001. [news summary or transcript link]

⁴ Channel 7 Today Tonight, SA. August 2006. [news coverage]

⁵ NSW Professional Standards Committee findings. 2007. [court/tribunal decision]

⁶ BBC News. Kathleen Folbigg: Misogyny helped jail her, science freed her. https://www.bbc.com/news/world-australia-65810056

⁷ NSW Court news. March 2017: Ryan domestic violence case. [news or court report]

⁸ McClellan J. Criminal Appeal Court ruling (2012) as quoted in Judging Kathleen Folbigg. Inside Story.

⁹ Daily Telegraph. Bar Association objects to Tedeschi directive. December 2017.

¹⁰ ABC News Australia. Kathleen Folbigg's tragic life started long before her babies died. https://www.abc.net.au/news/2023-06-03/kathleen-folbigg-life-story-before-wrongful-jailed-pardon/102420716

[11] BBC News. Folbigg ex-husband opposed to exoneration. https://www.bbc.com/news/world-australia-65810056

[12] Exoneration background materials. 2022–23 Inquiry. [official docs]

[13] Folbigg trial transcripts; Craig Folbigg statements. Various dates.

[14] The Conversation. SIDS Law brings grief, not clarity. https://theconversation.com/kathleen-folbiggs-children-likely-died-of-natural-causes-not-murder-156762

[15] 2019 Folbigg Inquiry. Expert Reports. https://2019folbigginquiry.dcj.nsw.gov.au/

[16] Interview with Dr. Janice Ophoven, Peter Berry; SIDS Law criticisms. [media, expert analysis]

[17] Inquiry exhibits: Garbutt report. [Folbigg Inquiry site]

[18] The Age. October 23, 2003; Foster mother letter reporting

[19] Busuttil expert opinion, cited in 2019 Folbigg Inquiry docs.

[20] Rulings of Justice Wood, Court of Appeals. February 2003. (public record)

[21] New South Wales Court of Criminal Appeal in R v Folbigg NSWCCA 23

[22] ANU Reporter. Freeing Kathleen Folbigg: genetics, truth and justice. https://reporter.anu.edu.au/all-stories/freeing-kathleen-folbigg-genetics-truth-and-justice

CHAPTER 11

[1] ABC News Australia. Kathleen Folbigg's tragic life started long before her babies died. https://www.abc.net.au/news/2023-06-03/kathleen-folbigg-life-story-before-wrongful-jailed-pardon/102420716

[2] 2019 Folbigg Inquiry. 2003 transcripts, Police surveillance tapes, disclosure timing.

https://2019folbigginquiry.dcj.nsw.gov.au/

[3] The Conversation. Kathleen Folbigg's children likely died of natural causes, not murder. https://theconversation.com/kathleen-folbiggs-children-likely-died-of-natural-causes-not-murder-156762

[4] Royal Statistical Society (RSS). *Statement on the misuse of statistics in the Sally Clark*

case. London: RSS; 2001. Available
from: https://rss.org.uk/miscfiles/clark.pdf

[5] Justice for Kathleen Folbigg. Forceps birth and
respiratory distress—2019 Inquiry exhibits.

[6] Blanch TR. Report of the Inquiry into the
Convictions of Kathleen Megan Folbigg. Sydney:
NSW Department of Communities and Justice;
2019. Available from:
https://www.justice.nsw.gov.au/justicepolicy/Docu
ments/folbigg-inquiry-report-2019.pdf

[7] Kennedy L, Cubby B. Folbigg guilty of killing her
four children. Sydney Morning Herald. 22 May
2003. Available from:
https://www.smh.com.au/national/folbigg-guilty-of-
killing-her-four-children-20030522-gdgqke.html

[8] ABC News. Folbigg guilty of killing her four
children. ABC News. 21 May 2003. Available from:
https://www.abc.net.au/news/2003-05-21/folbigg-
guilty-of-killing-her-four-children/1856484

[9] The Daily Telegraph. Jury convicts Folbigg of
killing her four children. The Daily Telegraph. 22
May 2003. Available from:
https://www.dailytelegraph.com.au/news/nsw/folbi

gg-verdict-2003/news-
story/0e1d07be7fe71689281e4cf1eff444e2

[10] R v Folbigg [2005] NSWCCA 23. Supreme
Court of New South Wales, Court of Criminal
Appeal; 17 February 2005. Available from:
https://www.caselaw.nsw.gov.au/decision/549f85b
4263b0a4600b3c4f2

[11] 2019 Folbigg Inquiry. Appeals and sentencing
records.

[12] Science.org. How a geneticist led the effort to
free a mother convicted of killing her children.
https://www.science.org/content/article/how-
geneticist-led-effort-free-mother-convicted-killing-
her-children

CHAPTER 12

[1] Cunliffe E. Murder, Medicine and Motherhood.
Hart Publishing; 2011. Medical Law Review
summary: https://academic.oup.com/medlaw/articl
e/20/4/769/1488399

[2] Murder, Medicine and Motherhood, book
review. https://lpbr.net/2011/11/murder-medicine-
and-motherhood.html

[3] Cummings H. Blood Vows. 2011. Newcastle Herald

coverage: https://newcastleherald.com.au/news/nsw/helen-cummings-profile

[4] Duff E. New science would let Folbigg go free. Sydney Morning Herald. 24 February 2013. Available from: https://www.smh.com.au/national/new-science-would-let-folbigg-go-free-20130223-2exu0.html

[5] Inverell Times. Folbigg family reacts to push for inquiry. Inverell Times. 28 February 2013. Available from: https://www.inverelltimes.com.au/story/1308258/folbigg-family-reacts-to-push-for-inquiry/

[6] Newcastle University. Fight for justice – Hippocampus, 2023. https://www.newcastle.edu.au/stories/2023/hippocampus/justice-for-folbigg

[7] Newcastle University. Legal Centre in push for judicial inquiry. https://www.newcastle.edu.au/news/story/justice-for-folbigg

[8] Cordner S. Report on the Folbigg deaths. 2015. 2019 Folbigg Inquiry Exhibit 2-Q: https://2022folbigginquiry.dcj.nsw.gov.au/documents/exhibit-2-q.pdf

9 Visentin L. Folbigg's lawyer slams delay over review of convictions. Sydney Morning Herald. 21 March 2021. Available from: https://www.smh.com.au/national/nsw/folbigg-s-lawyer-slams-delay-over-review-of-convictions-20210321-p57cm7.html

10 IMDb. Australian Story – From Behind Bars. 2018. https://www.imdb.com/title/tt9418044/

CHAPTER 13

1 ABC News Australia. Speakman orders Folbigg Inquiry. 2018. https://www.abc.net.au/news/2018-08-22/

2 2019 Folbigg Inquiry. Family history & DNA evidence. https://2019folbigginquiry.dcj.nsw.gov.au/

3 2019 Folbigg Inquiry. Genetic evidence debate. [PDF] https://2019folbigginquiry.dcj.nsw.gov.au/

4 Professor Schwartz letter, Exhibit 8, 2022 Folbigg Inquiry. https://2022folbigginquiry.dcj.nsw.gov.au/

5 2019 Folbigg Inquiry. Canberra Team, Crotti et al. European Heart Journal.

2019. https://academic.oup.com/eurheartj/article-abstract/40/30/2369/5486241

[6] Ovid. Calmodulin, sudden death, and the Folbigg case: genes in court. https://journals.lww.com/ovid/Fulltext/2024/05200/Calmodulin,_sudden_death,_and_the_Folbigg_case_.5.aspx

[7] 2019 Folbigg Inquiry. Final Report, Blanch KC. https://2019folbigginquiry.dcj.nsw.gov.au/

[8] Clancy R. The Australian; Immunologist critical of Inquiry cross-examination. https://www.abc.net.au/news/2019-03-17/

[9] Byrne C. Scientist slams adversarial nature of Folbigg inquiry. ABC News. 27 April 2023. Available from: https://www.abc.net.au/news/2023-04-27/kathleen-folbigg-inquiry-scientist-carola-vinuesa/102272338

CHAPTER 14

[1] Overgaard MT, Nyegaard M, et al. Report on analysis of the Folbigg CALM2 G114R mutation—2022 Inquiry Exhibit 6. https://vbn.aau.dk/files/71995846/Report_CALM2_Folbigg_2022.pdf

[2] Brohus M, Arsov T, Nyegaard M, Schwartz PJ, et al. Infanticide vs. inherited cardiac arrhythmias. Europace. 2021;23(3):441–450. https://academic.oup.com/europace/article/23/3/441/6003670

[3] Brohus M, Arsov T, Wallace DA, et al. **Infanticide vs. inherited cardiac arrhythmias.** *EP Europace.* 2021;23(3):441–450. doi:10.1093/europace/euaa272. https://academic.oup.com/europace/article/23/3/441/5983835

[4] Crotti L, et al. Clinical presentation of calmodulin mutations: the International Calmodulinopathy Registry. Eur Heart J. 2023. https://academic.oup.com/eurheartj/article/44/36/3872/7283417

[5] Brohus M, Overgaard MT, et al. Calmodulin mutations affecting Gly114 impair binding to cardiac sodium channels. Hum Mol Genet. 2023. https://pubmed.ncbi.nlm.nih.gov/37526648/

[6] Vinuesa CG, Aypar U, Haviv I, Makunin I, Walker RL, Lioy DT, et al. Exome sequencing identifies novel variants in multiple genes in a family with sudden infant death syndrome. Eur J Hum Genet. 2021;29(12):1721–1731. Available from: https://www.nature.com/articles/s41431-021-00920-5

7 MedlinePlus. MYH6 gene. https://medlineplus.gov/genetics/gene/myh6/

8 PubMed. MYH6 and congenital heart disease. https://pubmed.ncbi.nlm.nih.gov/21131983/

9 StatPearls. Mucopolysaccharidosis Type II (Hunter's Syndrome). https://www.ncbi.nlm.nih.gov/books/NBK516214/

CHAPTER 15

1 Australian Academy of Science. Public Petition and statement. 2021. https://www.science.org.au/news-and-events/news-and-media-releases/nobel-laureates-and-leading-scientists-call-kathleen-folbigg-be-pardoned

2 BBC News. Kathleen Folbigg: Could science free Australian jailed for killing her children? March 2021. https://www.bbc.com/news/world-australia-56311293

3 McDermott Q. Experts say Kathleen Folbigg's diaries show no sign of guilt. The Australian. Oct 2021.

4 Folbigg v Attorney General of New South Wales [2021] NSWCA 31. Supreme Court of New South Wales, Court of Appeal; 24 March 2021. Available from: https://www.caselaw.nsw.gov.au/decision/1785d9f5f880447f796c3f1d

5 Australian Academy of Science. Scientists call for reform following Folbigg decision. Media release. 25 March 2021. Available from: https://www.science.org.au/news-and-events/news-and-media-releases/scientists-call-reform-following-folbigg-decision

6 Australian Academy of Science. Statement by Professor Carola Vinuesa on CALM2 variants in the Folbigg case. Canberra: Australian Academy of Science; 26 March 2021. Available from: https://www.science.org.au/news-and-events/news-and-media-releases/statement-professor-carola-vinuesa-calm2-variants-folbigg-case

7 Wilson J. Calls to free child killer Folbigg 'an insult'. The Daily Telegraph. 27 March 2021. Available from: https://www.dailytelegraph.com.au/news/nsw/calls-

to-free-child-killer-folbigg-an-insult/news-
story/3b1f76de2ec9e2a94ae2c74e01b8e021

[8] Visentin L. Kathleen Folbigg writes to NSW
Attorney-General pleading for justice. Sydney
Morning Herald. 4 March 2021. Available from:
https://www.smh.com.au/national/nsw/kathleen-
folbigg-writes-to-nsw-attorney-general-pleading-
for-justice-20210304-p5778m.html

[9] Floyd BJ, et al. Proactive Variant Effect Mapping
Aids Diagnosis in Pediatric Cardiac Arrest. Circ
Genom Precis Med.
2023. https://www.ahajournals.org/doi/10.1161/CIR
CGEN.122.004913

[10] SW Government. Attorney General orders
second inquiry into Folbigg convictions. Media
release. Sydney: NSW Department of
Communities and Justice; 18 May 2022. Available
from: https://www.dcj.nsw.gov.au/news-and-
media/media-releases/attorney-general-orders-
second-inquiry-into-folbigg-convictions.html

[11] Bathurst T. Report of the Second Inquiry into the
Convictions of Kathleen Megan Folbigg. Sydney:
NSW Office of Special Inquiries; 2023. Available

from: https://www.nsw.gov.au/nsw-government/projects-and-initiatives/folbigg-inquiry

[12] Kato K, et al. Mild to sudden-lethal LQTS calmodulinopathies due to variable K+ and Ca2+ channel perturbation by N138K-CaM. J Clin Invest. 2022.

CHAPTER 16

[1] Bathurst T. Opening Address, Second Inquiry. 2022.

[2] Australian Academy of Science. Science has been heard at the Kathleen Folbigg Inquiry. 2023. https://www.science.org.au/news-and-events/news-and-media-releases/science-has-been-heard-at-the-kathleen-folbigg-inquiry

[3] Overgaard MT, Chen SRW, Makunin I, Vinuesa CG. Functional assessment of CALM2 variants (G114R, G114W) associated with sudden death in the Folbigg family. Evidence to the Second Inquiry into the Convictions of Kathleen Folbigg. Sydney: NSW Office of Special Inquiries; 2023. Available from: https://www.nsw.gov.au/nsw-government/projects-and-initiatives/folbigg-inquiry

4 Chen S, et al. Physiological Genomics Identifies Genetic Modifiers of LQTS Type 2 Severity. JCI. 2018. Report to Inquiry, Feb 2023.

5 Skotte L, et al. BSN gene variants linked to febrile seizures. Brain. 2022.
BMJ JMG. Variants in BSN gene associated with epilepsy. 2023.

CHAPTER 17

1 The Weekend Australian. February 2023; 2022 Folbigg Inquiry transcripts.

2 MacRae CA. CALM2 G114R and sudden death: caution in assigning pathogenicity.
Europace. 2023;25(1):289–90. Available from: https://doi.org/10.1093/europace/euad001

3 Nyegaard M, Brohus M, Overgaard MT. Reply to: CALM2 G114R and sudden death: caution in assigning pathogenicity.
Europace. 2023;25(2):391–2. Available from: https://doi.org/10.1093/europace/euad021

4 Schwartz PJ. Expert report to the Second Inquiry into the Convictions of Kathleen Folbigg. Milan: IRCCS Istituto Auxologico Italiano – Department of Molecular Cardiology; 2022. Available from: https://www.nsw.gov.au/nsw-government/projects-and-initiatives/folbigg-inquiry

5 Schwartz PJ. Expert report to the Second Inquiry into the Convictions of Kathleen Folbigg. Milan: IRCCS Istituto Auxologico Italiano – Department of Molecular Cardiology; 2023. Available from: https://www.nsw.gov.au/nsw-government/projects-and-initiatives/folbigg-inquiry

6 Kang et al. Calmodulin variant research. PubMed reference (full
preprint/abstract). https://pubmed.ncbi.nlm.nih.gov/37138298/

7 *Inquiry into the* Convictions *of Kathleen Megan Folbigg, Transcripts of the Inquiry evidence of Prof Carola Vinuesa and Professor Peter Schwartz.* Available
at: https://2022folbigginquiry.dcj.nsw.gov.au/transcripts.html2022folbigginquiry.dcj.nsw+1

8 Sun Y, Zhang Y, Li P, Li Y, Wang J, Xu X, et al. Identification of BSN gene variants in children with epilepsy: Evidence for pathogenic involvement. Epilepsia Open. 2023;8(1):85–96. Available from: https://doi.org/10.1002/epi4.12719

9 Liu Z, Zhou L, Chen Y, Zhao X, Wu H, Zhang J, et al. Enrichment of BSN variants among pediatric

patients with febrile seizures and epilepsy. Front Genet. 2023;14:1187456. Available from: https://doi.org/10.3389/fgene.2023.1187456

[10] Dibbens L. Expert report to the Second Inquiry into the Convictions of Kathleen Folbigg. Adelaide: University of South Australia; 2023. Available from: https://www.nsw.gov.au/nsw-government/projects-and-initiatives/folbigg-inquiry

[11] Dehorter N. Expert report to the Second Inquiry into the Convictions of Kathleen Folbigg. Canberra: Australian National University; 2023. Available from: https://www.nsw.gov.au/nsw-government/projects-and-initiatives/folbigg-inquiry

[12] Vinuesa CG, Makunin I, Aypar U, Haviv I, Lioy DT, Overgaard MT, et al. Genomic reanalysis report submitted to the Second Inquiry into the Convictions of Kathleen Folbigg. Canberra: Australian National University and NSW Office of Special Inquiries; 2023. Available from: https://www.nsw.gov.au/nsw-government/projects-and-initiatives/folbigg-inquiry

[13] Byard RW, Krous HF, Fleming PJ, et al. Suffocation, smothering, and sudden infant deaths: A review of current issues. Forensic

Sci Med Pathol. 2019;15(1):5–14. Available from:
https://doi.org/10.1007/s12024-018-0052-3

[14] Ryan M. Expert report to the Second Inquiry into
the Convictions of Kathleen Folbigg. Melbourne:
Royal Children's
Hospital and University of Melbourne; 2023.
Available from: https://www.nsw.gov.au/nsw-
government/projects-and-initiatives/folbigg-inquiry

[15] Bathurst T. Report of the Second Inquiry into the
Convictions of Kathleen Megan Folbigg. Sydney:
NSW Office of Special Inquiries; 2023. Available
from: https://www.nsw.gov.au/nsw-
government/projects-and-initiatives/folbigg-inquiry

[16] Bathurst T. Preliminary findings of the Second
Inquiry into the Convictions of
Kathleen Megan Folbigg. Sydney:
NSW Office of Special Inquiries; 1 June 2023.
Available from: https://www.nsw.gov.au/nsw-
government/projects-and-initiatives/folbigg-inquiry

[17] Governor pardons Kathleen Folbigg following
Bathurst KC report. Media release. Sydney:
Department of Communities and Justice; 5 June 20
23. Available from: https://www.nsw.gov.au/media-
releases/governor-pardons-kathleen-folbigg

18 Folbigg v R [2023] NSWCCA 295.
Supreme Court of New South Wales, Court of Crimi
nal Appeal; 14 December 2023. Available from:
https://www.caselaw.nsw.gov.au/decision/18ff4076
c566e86b75cd6f23

19 ABC News. Kathleen Folbigg speaks after her
acquittal: 'Today is a victory for science and truth'.
ABC News. 14 December 2023. Available from:
https://www.abc.net.au/news/2023-12-14/kathleen-
folbigg-convictions-quashed-speaks-
out/103237706

CHAPTE18

1 Australian Academy of Science. Kathleen Folbigg
Inquiry: Scientific evidence submission. Canberra:
AAS;   2022.
(https://www.science.org.au/supporting-
science/science-policy-and-
analysis/submissions/kathleen-folbigg-inquiry)

2 UK Parliament. Draft Criminal Justice (Science
Advisory   Panels)   Bill.   Autumn   2025.
(https://bills.parliament.uk/)

3 California Innocence Project. Science advisory panel proposals in US states. 2025. (https://californiainnocenceproject.org/)

4 Innocence Network. Innocence review commissions in the United States and United Kingdom. 2025. (https://innocencenetwork.org/innocence-commissions/)

5 State v. Nieves; State v. Cifelli, A-26/27-23. Supreme Court of New Jersey. November 20, 2025. (https://law.justia.com/cases/new-jersey/supreme-court/2025/a-26-27-23.html

6 State v. Darryl Nieves; State v. Michael Cifelli, A-26/27-23 (088683), Supreme Court of New Jersey, 19 November 2025.

: https://www.njcourts.gov/system/files/court-opinions/2025/a_26_27_23.pdf

7 Gross SR, O'Brien B, Hu C, Kennedy EH. Rate of false conviction of criminal defendants who are sentenced to death. Proc Natl Acad Sci USA. 2014;111(20):7230–7235. (https://www.pnas.org/doi/10.1073/pnas.13064171 11

8 National Registry of Exonerations. Exonerations in 2024 report. University of California Irvine Newkirk Center for Science & ociety; 2025.

(https://www.law.umich.edu/special/exoneration/Pages/about.aspx

---

CHAPTER 18

[1] Caffey J. Multiple fractures in the long bones of infants suffering from chronic subdural hematoma. Am J Roentgenol Radium Ther. 1946;56:163–173.

[2] Kempe CH, Silverman FN, Steele BF, et al. The battered-child syndrome. JAMA. 1962;181(1):17–24.

[3] Guthkelch AN. Infantile subdural haematoma and its relationship to whiplash injuries. BMJ. 1971;2(5759):430–431.

[4] Caffey J. On the theory and practice of shaking infants: its potential residual effects of permanent brain damage and mental retardation. Am J Dis Child. 1972;124(2):161–169.

[5] Ludwig S, Warman M. Shaken Baby Syndrome: A Review of 20 Cases. Ann Emerg Med. 1984;13(2):104–113.

[6] Duhaime AC, et al. The shaken baby syndrome: A clinical, pathological, and biomechanical study. J Neurosurg. 1987;66(3):409–415.

[7] Mikva A. US Court of Appeals, remarks on child abuse trials. 1990.

[8] Carty H, Ratcliffe J. Handle With Care leaflet. BMJ. 1995.

[9] Geddes J, et al. Dural haemorrhage in non-traumatic infant deaths. Neuropathol Appl Neurobiol. 2003;29(1):14–22.
Geddes JF, et al. The evidence base for shaken baby syndrome: We need to question the diagnostic criteria (Geddes I & II). Brain. 2001;124(5):961–969.

[10] National Association of Medical Examiners. SBS position paper. 2001. (Withdrawn 2006.)

[11] Donohoe M. Evidence-Based Medicine and Shaken Baby Syndrome. Am J Forensic Med Pathol. 2003;24(3):239–242.

[12] Geddes JF, Hackshaw AK, Vowles GH, et al. Shaken baby syndrome: anatomical and biomechanical perspectives (Geddes III). Neuropathol Appl Neurobiol. 2003;29(1):14–22.

[13] Ayoub D, et al. After the Court of Appeal: R v Harris and others EWCA Crim 1980. *Medicine, Science and the Law*. 2005;45(4):327–335.

[14] Findley KA, et al. Shaken baby syndrome, abusive head trauma, and actual innocence: getting it right. *Houston Journal of Health Law & Policy*. 2012;12(2):209–312.

[15] Squire W, Cohen M, Scheimberg I. GMC hearings and legal challenges. 2007–2014.

[16] American Academy of Pediatrics. Statement on Abusive Head Trauma. Pediatrics. 2009;123(5):1409–1411.

[17] Findley K. Oklahoma City University School of Law SBS Symposium. 2012.

[18] Judge Matthew Kennelly, Jennifer Del Prete release. US District Court, Chicago. 2014.

19.Squier v General Medical Council EWHC 2739 (Admin).judiciary+1

[20] Findley et al., "Feigned Consensus: Usurping the Law in Shaken Baby Syndrome," 2020

https://repository.law.umich.edu/cgi/viewcontent.cgi?article=3104&context=articles

---

APPENDIX II

[1] Froggatt P, et al. Multiple cases of sudden infant death syndrome in families. Br Med J. 1971;1(5741):321-324.

[2] Peterson DR, et al. Sudden infant death syndrome: recurrence in families. Pediatrics. 1980;65(5):1029-1032.

3 Jeffrey H, et al. Sudden infant death in families: Observations and review. Arch Dis Child. 1983;58(5):368-372.

4 Rosen CL. Recurrence of sudden infant death syndrome in families. S Afr Med J. 1983;63:184-185.

5 Diamond DG. Familial recurrence of sudden infant death: Case report. Pediatrics. 1986;78(4):812-814.

6 Emery JL. Multiple cases in the same family. Br J Paediatr. 1986;108(2):230-233.

7 Oren J, et al. Familial occurrence of SIDS. J Pediatr. 1987;110(3):473-478.

8 Beal SM, Blundell N. Epidemiology of SIDS recurrence in Australia: A review. Med J Aust. 1988;148:118-123.

9 Morild I, et al. SIDS recurrence rates in Scandinavian population groups. Scand J Infect Dis. 1989;21(4):337-340.

10 Guneroth B, et al. Twin studies and SIDS family recurrence. Eur J Pediatr. 1990;149(1):13-16.

[11] Wolkind S, et al. SIDS: An epidemiological study of recurrence in the UK. Community Med. 1993;15(3):193-198.

[12] CONI (Care of Next Infant) Project Group. National surveillance of families at risk. Arch Dis Child. 1998;78(3):209-215.

[13] Calenda E, et al. Familial SIDS: Review of cases in Italy. Ital J Pediatr. 2000;26(5):194-198.

[14] McFarland J, et al. Case report: Fifth occurrence of sudden infant death in a single family. Pediatrics. 2002;110(3):e45

[15] EagleSmith RE, et al. Indigenous American populations and SIDS recurrence statistics. Pediatrics. 2006;117(4):e758-e765.

[16] Garstang J, et al. Multiple infant deaths in families: UK prospective case review. Arch Dis Child. 2020;105(6):579-584.

www.ingramcontent.com/pod-product-compliance
Lightning Source LLC
Chambersburg PA
CBHW050026040726
47599CB00015B/1549